ANALOG COMMUNICATION ENGINEERING FUNDAMENTALS

Mainak Mukhopadhyay

Contents

Chapter 1

FOURIER ANALYSIS

FOURIER ANALYSIS

Analysis of periodic waveforms	Analysis of Aperiodic Waveform
(FOURIER SERIES)	(FOURIER TRANSFORM)
1.Analysis of discrete signal	1.Analysis of discrete signal
2. Analysis of continuous signal	2. Analysis of continuous signal

Introduction:-

Case 1----- when the sinusoidal wave of same frequency are added ,the resultant is also a sinusoidal of same FREQUENCY, only the AMPLITUDE and PHASE of the RESULTANT may be different.

Pictorial description:-

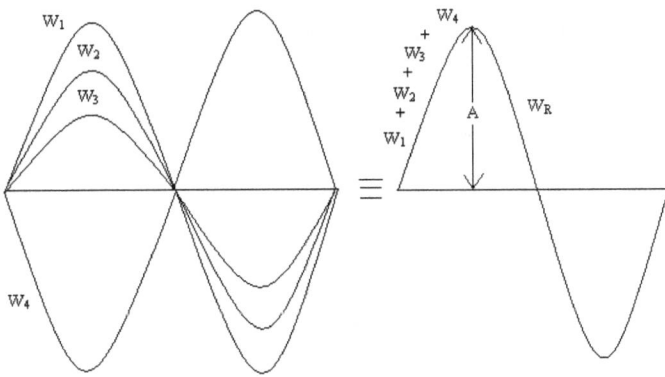

In the above example we added four sinusoidal signal of same frequency 'f' but they differ in amplitude say A_1, A_2, A_3 & A_4

Analog Communication Engineering Fundamentals

$W_R = W_1+W_2+W_3+W_4=(A_1+A_2+A_3+A_4)\sin\omega t$ i.e the resultant is also a sinusoid of same frequency but different amplitude

In the above example we eliminated the phase .Let us consider the more general case.
$W_1=A_1\sin(\omega t+_1)$, $W_2=A_2\sin(\omega t+_2)$, $W_3=A_3\sin(\omega t+_3)$, $W_4=A_4\sin(\omega t+_4)$------if we add those four signals when $A_1 \neq A_2 \neq A_3 \neq A_4$ & $\theta_1 \neq \theta_2 \neq \theta_3 \neq \theta_4$
We get------
$Wr=W_1+W_2+W_3+W_4=[A_1\cos\theta_1+A_2\cos\theta_2+A_3\cos_3+A_4\cos\theta_4]\sin\omega t+[A_1\sin\theta_1+A_2\sin\theta_2+A_3\sin\theta_3+A_4\sin\theta_4]\cos\omega t$
$=P\sin\omega t+Q\cos\omega t$
$=A\sin(\omega t+\theta)$
where $A=\sqrt{P^2+Q^2}$ & $\theta=\tan^{-1}Q/P\rightarrow$
where $P=\sum_{r=1}^{4}A_r\cos\theta_r$ & $Q=\sum_{r=1}^{4}A_r\sin\theta r$
So here resultant is also a sinusoid of same frequency but different amplitude and phase.

Case 2---- when a sinusoidal wave of different frequency are added then the resultant is not a sinusoidal waveform. and Situation A]$\rightarrow$ if the frequency ratio of the added harmonic waves to the fundamental wave is an INTEGER then the resultant is PERIODIC .Situation B]$\rightarrow$ if the above ratio is fraction then the resultant is an APERIODIC wave.

ANALYSIS:-
Let $X_1(t)=\sin\omega t$, and Let $X_1(t)$ is periodic with time period of T .i.e. $X_1(t+T)=X_1(t)$
So, obviously $\sin\omega(t+T)=\sin\omega t$ hence $\omega T=2\pi$ and we know $\cos n2\pi=1$ & $\sin n2\pi=0$.
Let $Xr(t)=\sin r\omega t$ where r is always an INTEGER now if we define –

$X(t)=\sum_{r=1}^{n}\sin r\omega t$,obviously lowest frequency in x(t) is for $X_1(t)$ which is ω and we know
$\omega T=2\pi$ i.e. $X_1(t)$ has time period of T as previously said
Now $X(t+T)=\sum_{r=1}^{n}\sin r\omega(t+T)$

$\qquad =\sum_{r=1}^{n}[\sin r\omega t\cos r\omega t+\sin r\omega t\cos r\omega t]$

$\qquad =\sum_{r=1}^{n}[\sin r\omega t\cos r2\pi+\sin r2\pi\cos r\omega t]$

$\qquad$ Since r is an integer$\rightarrow$

$\qquad =\sum_{r=1}^{n}\sin r\omega t =X(t)$

So,from the above analysis we get if a sinusoidal wave (Here x(t))is periodic with a period of T and if we added any no.of sinusoidal with that sinusoided with frequency integer multiple of that sinusoidal(i.e. of X(t)) then the resultant [X(t)] is also PERIODIC ,obviously resultant is not sinusoidal one .That completes the proof of SITUATION-A.
Now in the above discussion what happens if r is not always an INTEGER? Let us consider the case when r =F i.e.$X_F(t)=\sin F\omega t$ where F is a fraction not an integer. Assume $n_1<F<n_2$ where n_1 &n_2 are two integers.

Now obviously $X(t) = \sum_{r=1}^{n1} \sin r\omega t + \sin F\omega t + \sum_{r=n2}^{n} \sin r\omega t$

Now $X(t+T) = \sum_{r=1}^{n1} \sin r\omega(t+T) + \sin F\omega(t+T) + \sum_{r=n2}^{n} \sin r\omega(t+T)$

$$= \sum_{r=1}^{n1} \sin r\omega t + \sum_{r=n2}^{n} \sin r\omega t + \sin F\omega(t+T)$$

$\neq X(t)$; since F is a fraction

$\sin F\omega(t+T) \neq \sin F\omega t$

so now X(t) i.e resultant is not a periodic .So if we add at least one or more than one frequency which is not integer multiple of MINIMUM frequency(i.e. $X_1(t)$) then the resultant is neither SINUSOIDAL nor PERIODIC. That completes the proof of SITUATION B.

FOURIER SERIES

(ANALYSIS OF PERIODIC WAVEFORM)
In the year 1822,French mathematician J.B.J Fourier in the course of study of problem of heat flow showed that an arbitrary periodic function may be represented by an infinite series of sinusoids of harmonically related frequencies.

(Harmonically related frequency = frequency which is Integer multiple of fundamental frequency)

and we discussed previously that if we add harmonically related frequencies with fundamental frequency the resultant is also a periodic wave. So the Fourier's statement is just the REVERSE.

So any periodic function X(t) may be described by the following FOURIER SERIES→

$X(t) = A_0 + A_1 \cos\omega t + A_2 \cos 2\omega t + \ldots\ldots\ldots + A_n \cos n\omega t + \ldots\ldots$
$+ B_1 \sin\omega t + B_2 \sin 2\omega t + \ldots\ldots\ldots + B_n \sin n\omega t + \ldots\ldots$

Validity ? i.e. why both $\cos\omega t$ and $\sin\omega t$ are considered ? Why not PHASE shifted version is taken here? is it mathematically true to consider sinusoid treatment without phase?
J.B.J. Fourier and his followers does not provide any explicitly given answers for the above questions which would immediately born in a sensitive mind .

The original representation is

$$X(t) = \sum_{n=1}^{\infty} C_n \cos(n\omega_0 t + \theta_n) + C_0$$

(n=integer) ($C_0 = A_0$)

which consider only 'cos' term and also consider the PHASE of each signal .The above expression says that a signal x(t) is summation of an Infinite Series of SINUSOID frequency where each frequency is an integer multiple of FUNDAMENTAL(ω_0) and each frequency term has its own phase difference (θ_n) from this point we will replace the notation 'n' by 'k' because n is uniquely preserved for DISCRETE TIME notation.

So $X(t) = \sum_{k=1}^{\infty} C_k \cos(k\omega_0 t + \theta_k) + C_0 \ldots\ldots\ldots\ldots\ldots\ldots(1)$

Analog Communication Engineering Fundamentals

Where $C_0=A_0=$DC component

The above diagram shows the infinite SUM condition converge towards the original square wave but the main sinusoid ,which has some time period T_0 as of the original square wave is the major back bone of the above square wave i.e. $1/T_0=f_0$ frequency component contents approximate $\frac{3}{4}$ th share of ORIGINAL square wave ,and so this frequency is called FUNDAMENTAL frequency .All other frequencies which are integer multiple of FUNDAMENTAL has significantly small contribution compared to fundamental. It was observed that addition up to 5th or 7^{th}harmonic almost reproduce the ORIGINAL SIGNAL, higher harmonics have very small amplitude, so their effect can be neglected.

In the equation, C_k is the rms value of amplitude of frequency component K ω_0 ---- in spectrum analyzer.

We will see more or less

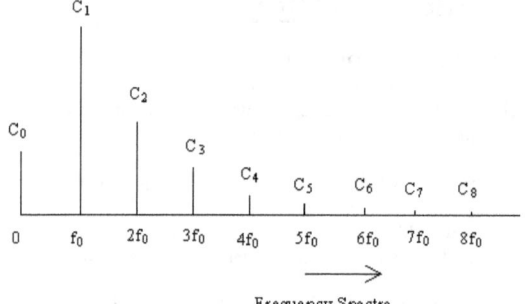

Frequency Spectra

similar diagram as shown above. Obviously C_1 is the fundamental amplitude. Now we have ----

$$X(t)=C_0+\sum_{k=1}^{\infty}C_k\cos(k\omega_0 t+\theta_k) \text{-----------------------------------}(1)$$

$$K= \text{Integer}$$
$$\omega_0 =\text{fundamental frequency}$$
$$C_0= \text{DC component}$$

Let $A_k=C_k\cos\theta_k$ & $B_k = C_k\sin\theta_k$

Obviously $\theta_k= \tan^{-1}\dfrac{B_k}{A_k}$ -------------------------------------- (2)

And $C_k=\sqrt{A_k^{2} + B_k^2}$ -------------------------------- (3)

But $C_0=A_0$

Now from (1)$\rightarrow$

$$X(t)=\sum_{k=1}^{\infty}[C_k\cos k\omega_0 t\cos\theta_k +C_k\sin k\omega_0 t\sin\theta_k]+C_0$$

$$= C_0 + \sum_{k=1}^{\infty}[A_k \cos k\omega_0 t + B_k \sin k\omega_0 t]$$

Replacing C_0 by A_0 we obtain the famous form of FOURIER SERIES–

$$X(t) = A_0 + \sum_{k=1}^{\infty}[A_k \cos k\omega_0 t + B_k \sin k\omega_0 t]$$

So now we can see where does 'PHASE' term vanish in above form and why does the 'sin' term appear. Its the 'sin' term which appears due to the phase difference in each frequency component. Note that ,if $\theta_k = 0$ i.e. there is no PHASE difference in any frequency component then $B_k=0$ i.e. the 'sin' term vanishes and also $A_k=C_k$; Then equation (4) reverts to equation (1) considering $\theta_k=0$ .At this point you probably got the answer why both 'sin' & 'cos' term appear in FOURIER SERIES, and that proves the validity of FOURIER SERIES form of equation (4).If we consider 'sin' instead of 'cos' in the equation(1) then it gives us same equation(4) .Just A_k & B_k will interchange their position .

Now from equation (1) we know that C_k is the amplitude of respective frequency component of $k\omega_0$ in frequency spectra to know the frequency spectra of a signal completely. We have to determine the C_k for each know we know $C_k = \sqrt{A_k^2 + B_k^2}$.

So if we are able to determine the value of A_k & B_k then it is easy to compute C_k

DETERMINATION OF A_0 ,A_k&B_k:-

In the Fourier series representation $\cos k\omega_0 t$ and $\sin k\omega_0 t$ is called a BASIS FUNCTION. These BASIS FUNCTION form an ORTHOGONAL set over the interval T_0. In that they satisfy the following set of relations.:-

1. $\displaystyle\int_{-T_0/2}^{+T_0/2} \cos(m\omega_0 t)\cos(k\omega_0 t)dt = [T_0/2 \qquad m=k$

$$[0 \qquad m \neq k$$

2. $\displaystyle\int_{-T_0/2}^{+T_0/2} \cos(m\omega_0 t)\sin(k\omega_0 t)dt = 0$

3. $\displaystyle\int_{-T_0/2}^{+T_0/2} \sin(m\omega_0 t)\sin(k\omega_0 t)dt = [T_0/2 \qquad m=k$

$$[0 \qquad m \neq k$$

now we know -----

$$X(t) = A_0 + \sum_{k=1}^{\infty}[A_k \cos k\omega_0 t + B_k \sin k\omega_0 t]$$

Analog Communication Engineering Fundamentals

Determination of $A_0 \rightarrow$
Integration on both side of (4) from $-T_0/2$ to $+T_0/2$

$$X(t)dt = A_0 \int_{-T_0/2}^{+T_0/2} dt + \sum_{k=1}^{\infty} [A_k \int_{-T_0/2}^{+T_0/2} \cos k\omega_0 t + B_k \int_{-T_0/2}^{+T_0/2} \sin k\omega_0 t]$$

$$= A_0 T_0 + 0 + 0$$

$$\because \int_{-T_0/2}^{+T_0/2} \cos k\omega_0 t \, dt = 0$$

$$\& \int_{-T_0/2}^{+T_0/2} \sin k\omega_0 t \, dt = 0$$

$$\therefore A_0 = \frac{1}{T_0} \int_{-T_0/2}^{+T_0/2} X(t) dt$$

Determination of A_k:-
Multiplying (4) by $\cos m\omega_0 t$ and integrating over the range $-T_0/2$ to $+T_0/2$ we get

$$\int_{-T_0/2}^{+T_0/2} X(t) \cos m\omega_0 t \, dt = \int_{-T_0/2}^{+T_0/2} A_0 \cos m\omega_0 t \, dt$$

$$+ \sum_{k=1}^{\infty} [A_k \int_{-T_0/2}^{+T_0/2} \cos k\omega_0 t \cos m\omega_0 t \, dt + B_k \int_{-T_0/2}^{+T_0/2} \cos m\omega_0 t \sin k\omega_0 t \, dt]$$

$$\text{or} \int_{-T_0/2}^{+T_0/2} X(t) \cos m\omega_0 t \, dt = 0 + A_m \frac{T_0}{2} + B_m 0$$

(when k=m)

$$= A_m \frac{T_0}{2} \text{ (when k=m)}$$

Writing k in place of m we get---

$$A_k = 2/T_0 \int_{-T_0/2}^{+T_0/2} X(t)\cos k\omega_0 t\, dt$$

Determination of B_k :-

Multiplying (4) by $\sin m\omega_0 t$ and then integrating over the range $-T_0/2$ to $+T_0/2$ of we have

$$\int_{-T_0/2}^{+T_0/2} X(t)\sin m\omega_0 t\, dt = \int_{-T_0/2}^{+T_0/2} A_0 \sin m\omega_0 t\, dt$$

$$+ \sum_{k=1}^{\infty}[A_k \int_{-T_0/2}^{+T_0/2}\cos k\omega_0 t\sin m\omega_0 t\, dt + B_k \int_{-T_0/2}^{+T_0/2}\sin m\omega_0 t\sin k\omega_0 t\, dt]$$

when m=k

$$= 0 + A_m + 0 + B_m$$

$$=. B_m \frac{T_0}{2}$$

Writing k in place of m we get

$$B_k = 2/T_0 \int_{-T_0/2}^{+T_0/2} X(t)\sin k\omega_0 t\, dt$$

So finally we get,

$$A_0 = \frac{1}{T_0}\int_{-T_0/2}^{+T_0/2} X(t)\, dt$$

$$A_k = 2/T_0 \int_{-T_0/2}^{+T_0/2} X(t)\cos k\omega_0 t\, dt$$

$$B_k = 2/T_0 \int_{-T_0/2}^{+T_0/2} X(t)\sin k\omega_0 t\, dt$$

The above formulas are EULER FORMULAS.

Let us define $a_0 = A_0 T_0$, $a_k = A_k T_0/2$ & $b_k = B_k T_0/2$ then another form of EULER FORMULA is

$$a_0 = \frac{1}{T_0} \int_{-T_0/2}^{+T_0/2} X(t)\, dt$$

$$a_k = 2/T_0 \int_{-T_0/2}^{+T_0/2} X(t)\cos k\omega_0 t\, dt$$

$$b_k = 2/T_0 \int_{-T_0/2}^{+T_0/2} X(t)\sin k\omega_0 t\, dt$$

we can use any one of the set according to our choice.
So the corresponding FOURIER SERIES relating to second set of EULER'S formula is

$$X(t) = a_0/T_0 + 2/T_0 \sum_{k=1}^{\infty} [A_k \cos k\omega_0 t + B_k \sin k\omega_0 t] \quad \text{--------------------------(6)}$$

We can use either eqn (4) or eqn (6).

ALTERNATIVE form of FOURIER SERIES:-

From (6) we know $C_0 = A_0$

Now $\cos k\omega_0 t = (e^{jk\omega_0 t} + e^{-jk\omega_0 t})/2$

$\sin k\omega_0 t = (e^{jk\omega_0 t} - e^{-jk\omega_0 t})/2j$

Substituting in eqn (6) ------------

$$X(t) = A_0 + 2/T_0 \sum_{k=1}^{\infty} [(\frac{a_k}{2} + \frac{b_k}{2})e^{jk\omega_0 t} + (\frac{a_k}{2} - \frac{b_k}{2})e^{-jk\omega_0 t}]$$

$$= A_0 + 1/T_0 \sum_{k=1}^{\infty} [(a_k - jb_k)e^{jk\omega_0 t} + (a_k + jb_k)e^{-jk\omega_0 t}]$$

Let $C_k = a_k - jb_k$ so $C_k^* = C_{-k} = a_k + jb_k$

So $X(t) = A_0 + 1/T_0 \sum_{k=1}^{\infty} (C_k e^{jk\omega_0 t} + C_{-k} e^{-jk\phi\omega_0 t})$

$$= 1/T_0 \sum_{k=-\infty}^{\infty} C_k e^{jk\omega_0 t}$$

$(C_0 = a_0)$

$$X(t) = 1/T_0 \sum_{k=-\infty}^{\infty} C_k e^{jk\omega_0 t}$$

-------(7)

This is the alternative form of Fourier series which is widely used in practical purpose. Here C_k is the amplitude of each frequency component.

We know $C_k = a_k - jb_k$

Analog Communication Engineering Fundamentals

$$= \int_{-T_0/2}^{+T_0/2} X(t)[\cos k\omega_0 t \quad j \sin k\omega_0 t]dt$$

$$= \int_{-T_0/2}^{+T_0/2} X(t)e^{-jk\omega_0 t} dt$$

$$C_k = \int_{-T_0/2}^{+T_0/2} X(t)e^{-jk\omega_0 t} dt$$

------------------(8)

$$C_0 = \int_{-T_0/2}^{+T_0/2} X(t)dt$$

So finally we have ----

$$X(t)=A_0 + \sum_{k=1}^{\infty}(A_k \cos k\omega_0 t - B_k \sin k\omega_0 t)$$

$$= a_0 \frac{a_0}{T_0} + \frac{2}{T_0}\sum_{k=1}^{\infty}(a_k \cos k\omega_0 t + b_k \sin k\omega_0 t)$$

where $a_0 = A_0 T_0$ $a_k = A_k T_0/2$ $b_k = B_k T_0/2$

here $a_0 = \int_{-T_0/2}^{T_0/2} x(t)dt$

$$a_k = \int_{-T_0/2}^{T_0/2} X(t)\cos k\beta\omega_0 t dt \qquad \& \quad b_k = \int_{-T_0/2}^{T_0/2} x(t)dt \sin k\omega_0 t \, dt$$

Alternatively ---------

$$X(t)=\frac{1}{T_0}\sum_{k=-\infty}^{\infty}C_k e^{jk\omega_0 t}$$

Where $C_k = \int_{-T_0/2}^{+T_0/2} X(t)e^{-jk\omega_0 t} dt$ -

$$\& \; C_0 = \int_{-T_0/2}^{T_0/2} x(t)dt \qquad \text{obviously } C_0 = a_0$$

DISCRETE TIME FOURIER SERIES:

So far we have considered only the continuous time Fourier series. Before proceeding further we have to realize what is discrete time frequency.

We know for continuous periodic signal f=1/T, where T is the time period. Now discrete signal is an ensemble of an of sampled values i.e discrete signal is a sampled version of a continuous signal. Since it is not continuous it is more convenient to represent it's periodicity by number of samples instead of time. Because all sampling points are on the time axis, number of samples is an indication of time period of a DISCRETE SIGNAL. So in similar manner we can represent the discrete time signal period as f=1/N, where N is the number of samples after which the signal will repeat.

Look at the similarity again-

For continuous signal-
T seconds = 1 period
So 1 second = 1/T period
and frequency = no.of period occur in 1 second = 1/T

For DISCRETE signal-
N samples = 1 period
So 1 sample = 1/N period
and frequency = no. of period occur in 1 sample = 1/N
Since $N \geq 1$ for a discrete time signal, frequency of a discrete time signal $f \leq 1$ (but it's not always true for continuous signal).

Now for periodic discrete time signal fundamental frequency $f_0 = 1/N_0$ or $\omega_0 = \dfrac{2\pi}{N_0}$ similar to continuous time.

Now in the span of 2π rad/samples frequency discrete signals are exist and due to the spacing of $2\pi/N_0$ (rad/sample)/sample there are $2\pi/(2\pi/N_0)$ samples or N_0 discrete samples are exist.So should be added over a range of K=<N_0>.

For continuous time we have

$$X(t) = \frac{1}{T_0} \sum_{k=-\infty}^{\infty} C_k e^{jk\omega_0 t}$$

Replacing t by n we have its discrete form

$$X[n]= \frac{1}{N_0} \sum_{k=\langle N_0\rangle} C_k e^{jk\omega_0 n}$$

----------------------------10)

k= $\langle N_0 \rangle$ denotes that k is taken over a range of N_0, it may be 0 to N_0-1 or 1 to N_0 or 2 to N_0+1..........or r to N_0+r-1 i.e. the total no of sample considered is N_0 only because after N_0 samples are over after

now $e^{jk\left(2\pi/N\right)n} = \cos[k\frac{2\pi}{N}n] + j\sin[k\frac{2\pi}{N}n]$

now $\cos(k\frac{2\pi}{N}n)=1$ if k=0,$\pm$N, $\pm$2N..........

(Since n is always integer)

if k$\neq$0, $\pm$N, $\pm$2N........then let

$$S=\sum_{n=0}^{N-1} e^{jk\left(2\pi/N\right)n} =\sum_{n=0}^{N-1}[e^{jk\left(2\pi/N\right)}]^n$$

Now $e^{jk\left(2\pi/N\right)}S = \sum_{n=0}^{N-1}[e^{jk\left(2\pi/N\right)}]^{n+1}$

$$= S-1+\sum_{n=0}^{N-1}[e^{jk\left(2\pi/N\right)}]^N$$

$$= S-1 +e^{jk2\pi}$$

[since k is an integer $e^{jk2\pi}=1$]

$$= S-1+1 =S$$

$\therefore$ s=0 (for k$\neq$0, $\pm$N, $\pm$2N..........)

So finally we get -------

$$\sum_{n=0}^{N-1}[e^{jk\left(2\pi/N\right)}]^n = \begin{cases} N for k = 0,\pm N,\pm 2N \\ 0 otherwise \end{cases}$$

$$\sum_{n=\langle N\rangle} e^{jk\left(2\pi/N\right)}n = \begin{cases} N for k = 0,\pm N,\pm 2N \\ 0 otherwise \end{cases}$$

.... (11)

Now multiplying eqn (10) by $e^{-jr\left(2\pi/N\right)n}$ and summing over n terms we obtain

$$\sum_{n=\langle N\rangle} X[n]\ e^{-jr\left(2\pi/N\right)n} = \sum_{n=\langle N\rangle}\sum_{k=\langle N\rangle} C_k e^{j(k-r)\ (2\pi/N)n}$$

$$= \sum_{n=\langle N \rangle} C_k \sum_{k=\langle N \rangle} e^{j(k-r)\,(2\pi/N)n}$$

now if k=r then $\sum_{n=\langle N \rangle} e^{j(k-r)\,(2\pi/N)n} = N$

$= 0$ otherwise

(i.e. when $K \neq r$)

(According to eqn 11)

so $\sum_{n=\langle N \rangle} X[n]\; e^{-jr\left(2\pi/N\right)n} = \sum_{k=\langle N \rangle} C_r N = C_r N$

or $C_r = 1/N \sum_{n=\langle N \rangle} X[n]\; e^{-jr\left(2\pi/N\right)n}$

replacing r by k and N by N_0 we get

$C_k = 1/N_0 \sum_{n=\langle N \rangle} X[n]\; e^{-jr\left(2\pi/N\right)n}$.

So the discrete time Fourier series is --------

$$X[n] = \sum_{k=\langle N_0 \rangle} C_k e^{jk\omega n_0} = \sum_{k=\langle N_0 \rangle} C_k e^{jk\left(2\pi/N_0\right)n}$$

And $C_k = 1/N_0 \sum_{n=\langle N \rangle} X[n]\; e^{-jr\left(2\pi/N\right)n}$.

$= 1/N_0 \sum_{n=\langle N \rangle} X[n]\; e^{-jr\left(2\pi/N_0\right)n}$

-------------------------------- (10) & (12)

PROPERTIES OF FOURIER SERIES

Both the continuous and discrete time Fourier series have some general properties:-

CONTINUOUS TIME FOURIER SERIES	DISCRETE TIME FOURIER SERIES
1.Linearity:- if $X(t) \leftrightarrow a_k$ $y(t) \leftrightarrow b_k$ And if $Z(t)=AX(t)+By(t)$ And if $Z(t) \leftrightarrow C_k$ Then $C_k= Aa_k+Bb_k$ 2.Time shifting :- if $X(t) \leftrightarrow a_k$ then $X(t-t_0) \leftrightarrow a_k e^{-jk\omega_0 t_0}$ $\rightarrow$ Hence $X(t)= \sum_{k=-\infty}^{\infty} a_k e^{jk_0 t_0}$ $f[X(t-t_0)]=1/T_0 \int_{-T_0/2}^{+T_0/2} X(t-t_0)e^{-jk\omega_0 t} dt$ let $t-t_0=\tau$ hence $d\tau /dt=1$ $=1/T_0 \int_{-T_0/2 -t_0}^{+T_0/2 -t_0} X(\tau)e^{-jk\omega_0 (t_0+\tau)} d\tau$ $= 1/T_0 \int_{T_0} X(\tau)e^{-jk\omega_0 \tau} . e^{-jk\omega_0 t_0} d\tau$ $=e^{-jk\omega_0 t_0} \frac{1}{T_0} \int_{T_0} X(\tau)e^{-jk\omega_0 \tau} d\tau$ $=a_k . e^{-jk\omega_0 t_0}$ $\omega_0 = \frac{2\pi}{T_0}$ 3.Frequency shift:- if we multiply any signal $X(t)$ by $e^{jm\omega_0 t}$ then resultant is obviously a frequency shifted version of $X(t)$ because let us assume $X(t)=Ae^{-}$	1.Linearity:- if $X[n] \leftrightarrow a_k$ $y[n] \leftrightarrow b_k$ $Z[n]= AX[n]+By[n]$ And $Z[n] \leftrightarrow C_k$ Then $C_k=Aa_k+Bb_k$ 2.Time shifting:- if $X[n] \leftrightarrow a_k$ then $X[n-n_0] \leftrightarrow a_k e^{-jr(2\pi/N)n}_0$ $= a_k e^{jk\omega_0 n_0}$ $\rightarrow f[X(n-n_0)]=\frac{1}{N_0} \sum_{n=\langle N_0 \rangle} X[n-n_0]$ $= e^{-jk\omega_0 n_0}$ where $\omega_0 = \frac{2\pi}{N_0}$ now let $n-n_0=m$ so $f[X(n-n_0)]= 1/N_0 \sum_{m=\langle N_0 \rangle} X[m] e^{-jk\omega_0 m}$ $= a_k e^{-jk\omega_0 n_0}$ $\omega_0 = \frac{2\pi}{N_0}$ 3.Frequency shift:- $e^{jm\omega_0 n}X[n] \leftrightarrow a_{k-m}$

$jq\omega_0 t$

so $e^{jm\omega_0 t}$ $X(t)=Ae^{j(m+q)\omega_0 t}$.

Obviously $(m+q)\omega_0$ denotes the frequency shifting .now if $X(t) \leftrightarrow a_k$

Then $e^{jm\omega_0 t}$ $X(t)=a_{k-m}$

Where $\omega_0 = 2\pi/T_0$

Now $f[e^{jm\omega_0 t}$ $X(t)]=$

$1/T_0 \int\limits_{-T_0/2}^{+T_0/2} X(t)e^{-jk\omega_0 t}e^{jm\omega_0 t}dt$

$= 1/T_0 \int\limits_{T_0} X(t)e^{-j(k-m)\omega_0 t}dt$

$= a_{k-m}$

4.Conjugation:-

if $X(t) \leftrightarrow a_k$

then if $X^*(t) \leftrightarrow a^*_k$

$\rightarrow$ now $X(t)= 1/T_0 \sum\limits_{k=-\infty}^{\infty} a_k e^{jk\omega_0 t}$

$X^*(t)= 1/T_0 \sum\limits_{k=-\infty}^{\infty} a^*_k e^{-jk\omega_0 t}$

Writing –k in place of k$\rightarrow$

$X^*(t)= 1/T_0 \sum\limits_{k=-\infty}^{\infty} a^*_{-k} e^{-jk\omega_0 t}$

Hence $X^*(t) \leftrightarrow a^*_{-k}$

(proved)

5.Time reversal:-

if $X(t) \leftrightarrow a_k$

then if $X(-t) \leftrightarrow a_{-k}$

$\rightarrow X(t)= 1/T_0 \sum\limits_{k=-\infty}^{\infty} a_k e^{jk\omega_0 t}$

$\rightarrow$ we know $a_k=1/N_0 \sum\limits_{n=\langle N\rangle} X[n] e^{-jk\omega_0 n}$

$\omega_0 = 2\pi/N_0$

now $f(e^{-jk\omega_0 n} X[n])$

$=1/N_0 \sum\limits_{n=\langle N_0\rangle} X[n] e^{-j(k-m)\omega_0 n}$

a_{k-m}

4.Conjugation:-

if $X[n] \leftrightarrow a_k$

then $X^*[n] \leftrightarrow a^*_{-k}$

$\rightarrow a_k= 1/N_0 \sum\limits_{n=\langle N\rangle} X[n] e^{-jk\omega_0 n}$

where $\omega_0 = 2\pi/N_0$

$a^*_k = 1/N_0 \sum\limits_{n=\langle N\rangle} X^*[n] e^{jk\omega_0 n}$

writing –k in place of k$\rightarrow$

$a^*_k = 1/N_0 \sum\limits_{n=\langle N\rangle} X^*[n] e^{-jk\omega_0 n}$

hence $X^*[n] \leftrightarrow a^*_k$

(proved)

5.Time reversal:-

if $X[n] \leftrightarrow a_k$

then$X[-n] \leftrightarrow a_{-k}$

$$\omega_0 = \frac{2\pi}{T_0}$$

$$X(-t) = 1/T_0 \sum_{k=-\infty}^{\infty} a_k e^{-jk\omega_0 t}$$

Writing –k in place of k→

$$X(-t) = 1/T_0 \sum_{k=-\infty}^{\infty} a_{-k} e^{jk\omega_0 t}$$

Hence $X(t) \leftrightarrow a_{-k}$

6.Time scaling

now if we are scaling the signal X(t)by α we get X(αt) where α >0 .obviously X(α t) is periodic with time period T_0/α

now $f[X(t)] = 1/T_0 \int_{-T_0/2}^{T_0/2} X(t)e^{-jk\omega_0 t} dt$

$= a_k$

so $f[X(\alpha t)] = 1/T_0 \int_{-\alpha T_0/2}^{\alpha T_0/2} X(\alpha t)e^{-jk\omega_0(\alpha t)} d(\alpha t)$

let $\tau = \alpha t$

$= 1/T_0 \int_{-\tau_0/2}^{\tau_0/2} X(\tau)e^{-jk\omega_0 \tau t} d(\tau)$

$= a_k$

hence $X(\alpha t) \leftrightarrow a_k$

7.Periodic convolution:-
for two periodic signal X(t) & y(t)'s

$$a_k = 1/N_0 \sum_{n=\langle N\rangle} X[n]\, e^{-jk\omega_0 n}$$

where $\omega_0 = \frac{2\pi}{N_0}$

$$a_{-k} = 1/N_0 \sum_{n=\langle N_0\rangle} X[n]\, e^{jk\omega_0 n}$$

writing –n in place of n→

$$a_{-k} = 1/N_0 \sum_{n=\langle N_0\rangle} X[-n]\, e^{-jk\omega_0 n}$$

Hence $X[-n] \leftrightarrow a_{-k}$

6.Time scaling:-

let $X_m[n] = 1/m X[n/m] \sum_{n=\langle m\rangle} e^{jkn\left(\frac{2\pi}{m}\right)}$

note:- if m=1 and k& n are always integer (always hold for discrete signal)
then $X_m[n] = X[n]$ since

$$\sum_{n=\langle m\rangle} e^{jkn\left(\frac{2\pi}{m}\right)} = m$$

Now if n=0, $\pm$m, $\pm$2m..........

then $\sum_{n=\langle m\rangle} e^{jkn\left(\frac{2\pi}{m}\right)} = m$

$= 0$ otherwise.

So finally we have
$X_m[n] = X[n/m]$
    If n is an integer multiple
    $= 0$ otherwise
now $f(X[n/m]) =$

$$1/mN_0 \sum_{n=\langle N_0\rangle} X[n/m]\, e^{-jkn\left(\frac{2\pi}{mN_0}\right)}$$

let P=n/m
because if X[n] is periodic by N_0 then X[n/m] is periodic when $mN_0 = 1/m$

$.1/N_0 \sum_{n=\langle N_0\rangle} X[P]\, e^{-jkP\left(\frac{2\pi}{N_0}\right)} = 1/m\, a_k$

hence $X[n/m] \leftrightarrow 1/m.a_k$
if n is an integer multiple of m
otherwise $X[n/m] = 0$

convolution integral defined as X(t)*y(t)= $\int_{T_0} X(\tau)y(t-\tau)d\tau$ now $f[\int_{T_0} X(\tau)y(t-\tau)d\tau]$ $= 1/T_0 \int_{T_0}\int_{T_0} X(\tau)y(t-\tau)d\tau$ $= 1/T_0 \int_{T_0} X(\tau) \int_{T_0} \left[\sum_{k=-\infty}^{\infty} b_k e^{jk\omega_0(t-\tau)}\right] e^{-jk\omega_0 t} \, d\tau dt$ let us assume $k=k_1$ $= \int_{T_0} b_{k1}\left[1/T_0\int_{T_0} X(\tau)e^{-jk\omega_0\tau}d\tau\right]dt$ $= \int_{T_0} b_{k1}a_{k1}dt = T_0 b_{k1}a_{k1} = T_0 a_k b_k$ $(\because k_1=k)$ $\int_{T_0} X(\tau)y(t-\tau)d\tau = T_0 a_k b_k$ 8.<u>Multiplication</u>:- $f[X(t)y(t)]=1/T_0 \int_{T_0} X(t)y(t) \, e^{-jk\omega_0 t} dt$ now $X(t)= \sum_{l=-\infty}^{\infty} a_l e^{jk\omega_0 t}$ $= \sum_{l=-\infty}^{\infty} a_l (1/T_0 \int_{T_0} y(t)e^{j\omega_0(k-l)t} dt)$ $= \sum_{l=-\infty}^{\infty} a_l b_{k-l}$ $X(t)y(t) = \sum_{l=-\infty}^{\infty} a_l b_{k-l}$	7.<u>Periodic convolution</u>:- for two periodic signal X[n] & Y [n]'s convolution sum is defined as $X[n] *Y[n]= \sum_{r=\langle N_0\rangle} X[r] \, Y[n-r]$ Now $\sum_{r=\langle N_0\rangle} X[r] \, Y[n-r]= \sum_{r=\langle N_0\rangle} X[r] \sum_{k=\langle N_0\rangle} b_k e^{jk\omega_0(n-r)}$ $=N_0 \sum_{k=\langle N_0\rangle} b_k \left[1/N_0 \sum_{r=\langle N_0\rangle} X[r]e^{-jk\omega_0 r}\right] e^{jk\omega_0 n}$ $= \sum_{k=\langle N_0\rangle} N_0 a_k b_k e^{jkn\omega_0}$ $\sum_{r=\langle N_0\rangle} X[r] \, Y[n-r] \leftrightarrow N_0 a_k b_k$ 8.<u>Multiplication</u>:- $f([X[n]y[n])=1/N_0 \sum_{n=\langle N_0\rangle} X[n]Y[n]e^{-j\omega_0 m}$ now $X[n]= \sum_{l=\langle N_0\rangle} a_l e^{jl\omega_0 n}$ or $f([X[n]y[n])=1/N_0 \sum_{n=\langle N_0\rangle} \sum_{l=\langle N_0\rangle} a_l Y[n]e^{-j\omega_0(k-l)n}$ $= \sum_{l=\langle N_0\rangle} a_l \; 1/N_0 \sum_{n=\langle N_0\rangle} Y[n]e^{-j\omega_0(k-l)n}$ $= \sum_{l=\langle N_0\rangle} a_l b_{k-l}$

9.Differentiation:-

we know $X(t) = \sum_{k=-\infty}^{\infty} a_k e^{-jk\omega_0 t}$

So $dX(t)/dt = jk\omega_0 \sum_{k=-\infty}^{\infty} a_k e^{jk\omega_0 t} = jk\omega_0 X(t)$

So $f[dX(t)/dt] = f[jk\omega_0 X(t)] = jk\omega_0 f[X(t)]$

So $f[dX(t)/dt] = jk\omega_0 a_k$

$\therefore dX(t)/dt \leftrightarrow jk\omega_0 a_k$

$\therefore d^n X(t)/dt^n \leftrightarrow (jk\omega_0)^n a_k$

10.INTEGRATION:-

$x(t) = \dfrac{1}{T_0} \sum_{K=-\infty}^{+\infty} a_K e^{+jK\omega_0 t}$

$\int_{-\infty}^{t} x(t)dt = \dfrac{1}{jK\omega_0} \sum_{K=-\infty}^{+\infty} a_K (e^{+jK\omega_0 t})^{t}_{-\infty}$

$\int_{-\infty}^{t} x(t)dt = \dfrac{1}{jK\omega_0} \sum_{K=-\infty}^{+\infty} a_K e^{+jK\omega_0 t} = \left(\dfrac{1}{jK\omega_0}\right) x(t)$

$So f\left[\int_{-\infty}^{t} x(t)dt\right] = \left(\dfrac{1}{jK\omega_0}\right) f[x(t)] = \left(\dfrac{1}{jK\omega_0}\right) a_k$

$Therefore \int_{-\infty}^{t} x(t)dt \leftrightarrow \left(\dfrac{1}{jK\omega_0}\right) a_k$

$[X[n]y[n] \leftrightarrow \sum_{l=\langle N_0 \rangle}^{\infty} a_l b_{k-l}$

9.First difference :-

we know $X[n] = \sum_{k=\langle N_0 \rangle} a_k e^{jk\omega_0 n}$

and $X[n-1] = \sum_{k=\langle N_0 \rangle} a_k e^{jk\omega_0 (n-1)}$

so $X[n]-X[n-1] = \sum_{k=\langle N_0 \rangle} a_k e^{jk\omega_0 n} -$

$\sum_{k=\langle N_0 \rangle} a_k e^{jk\omega_0 (n-1)} e^{-jk\omega_0}$

$= (1- e^{-jk\omega_0}) \sum_{k=\langle N_0 \rangle} a_k e^{jk\omega_0 n}$

$= (1- e^{-jk\omega_0}) X[n]$

now $f(X[n]-X[n-1]) = f[(1- e^{-jk\omega_0}) X[n]]$

$= (1- e^{-jk\omega_0}) f[X[n]]$

$= (1- e^{-jk\omega_0}) a_k$

$X[n]-X[n-1] \leftrightarrow (1- e^{-jk\omega_0}) a_k$

10.Running Sum

$X[n] = \sum_{K=\langle N_0 \rangle} a_K e^{jK\omega_0 n}$

$So \sum_{n=-\infty}^{n} X[n] = \sum_{-\infty}^{n} \sum_{K=\langle N_0 \rangle} a_K e^{jK\omega_0 n}$

$\Rightarrow \sum_{n=-\infty}^{n} X[n] = a_{K_1} \sum_{n=-\infty}^{n} e^{+jK_1\omega_0 n}$

$\Rightarrow \sum_{n=-\infty}^{n} X[n] = a_{K_1} \dfrac{1}{1- e^{jK_1\omega_0}}$

ANALYSIS OF NON-PERIODIC PULSE TRAIN

Frequency spectra of periodic pulse train(FSPPT):-

Consider a periodic pulse train x(t) as

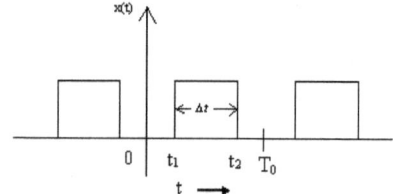

Where,

$$x(t) = 0 \quad 0 < t < t_1$$
$$\quad\quad = A \quad t_1 \le t \le t_2$$
$$\quad\quad = 0 \quad t_2 < t \le T_0$$

Period of PPT is T_0 , so $\omega_0 = \dfrac{2\pi}{T_0}$

From Fourier series representation we know

$$C_K = A \int_{t_1}^{t_2} e^{-jK\omega_0 t} \, dt = \frac{jA}{K\omega_0} \Big[e^{-jK\omega_0 t} \Big]_{t_1}^{t_2}$$

$$= \frac{ja}{K\omega_0} e^{-jk\omega_0 t_1} \Big[e^{-jK\omega_0 (t_2 - t_1)} - 1 \Big]$$

Let $\Delta t = t_2 - t_1$ =pulse width

$$= \frac{ja}{K\omega_0} e^{-jk\omega_0 t_1} \Big[e^{-jK\omega_0 \Delta t} - 1 \Big]$$

$$= -\frac{ja}{K\omega_0} e^{-jk\omega_0 t_1} e^{-jK\omega_0 \Delta t/2} \left[e^{-jK\omega_0 \Delta t/2} - e^{jK\omega_0 \Delta t/2} \right]$$

$$= \frac{A}{K\omega_0} e^{-jK\omega_0 (t_1 + \Delta t/2)} 2\left[e^{jK\omega_0 \Delta t/2} - e^{-jK\omega_0 \Delta t/2} \right]/2j$$

$$= \frac{2A}{K\omega_0} e^{-jK\omega_0 (t_1 + \Delta t/2)} \sin(K\omega_0 \Delta t/2)$$

$$= 2A. \frac{\Delta t}{2}. e^{-jK\omega_0 (t_1 + \Delta t/2)} \frac{\sin(K\omega_0 \Delta t/2)}{K\omega_0 \Delta t/2}$$

$$= A.\Delta t. \sin c(K\omega_0 \Delta t/2)$$

Let assume $\phi_K = K\omega_0 \left(t_1 + \frac{\Delta t}{2} \right)$

$$= A\Delta t \sin c(K\pi f_o \Delta t) e^{-j\phi_K}$$

So for FSPPT finally we get-

$$C_K = A\Delta t \sin c(K\pi f_o \Delta t) e^{-j\phi_K} \qquad\underline{\textbf{A}}$$

Where $\Delta t = t_2 - t_1$ =pulse width

and $\phi_K = K\omega_0 \left(t_1 + \frac{\Delta t}{2} \right)$

Substituting **A** in the Fourier series

$$x(t) = \frac{1}{T_0} \sum_{K=-\infty}^{+\infty} C_K e^{+jK\omega_0 t}$$

$$= \frac{A\Delta t}{T_0} \sum_{K=-\infty}^{\infty} \sin c(K\pi f_o \Delta t) e^{-j(K\omega_0 - \phi_K)}$$

$$So |C_K| = A\Delta t \sin c(K\pi f_o \Delta t)$$

$$\arg C_K = -\phi_K = -K\omega_0 (t_1 + \Delta t/2)$$

Now if $C_K = 0$

$$Sinc(K\pi f_0 \Delta t) = 0$$

$$\Rightarrow Sinc(K\pi f_0 \Delta t) = 0$$

Then $\Rightarrow (K\pi f_0 \Delta t) = r\pi$

$$\Rightarrow K f_0 = r/\Delta t$$

$$\Rightarrow K = r(T_0/\Delta t)$$

Let's enjoy some modification:-
Shift the PPT in the way that t_1 coincide with the origin. Obviously now
$\Delta t = t_2 - t_1 = \Delta t = t_2 = \tau$ (Let assume)
At this time time period is as usual T_0

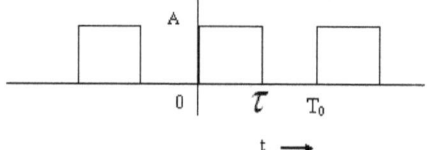

So $K = r(T_0 / \tau)$

Now when K=0 , $Sinc(K\pi f_0 \tau) = 1$

So at zero frequency $|C_0| = A_\tau$,obviously this is the D.C component.

Now $|C_K| = A\tau \sin c (K\pi f_0 \tau)$

So $\dfrac{d|C_k|}{d(K\pi f_0 \tau)} = A_\tau \left[\dfrac{\cos(K\pi f_0 \tau)}{(K\pi f_0 \tau)} - \dfrac{\sin(K\pi f_0 \tau)}{(K\pi f_0 \tau)^2} \right]$

Now at the maximum value of $|C_K|$, $\dfrac{d|C_K|}{d(K\pi f \tau_0)} = 0$

So $\tan(K\pi f_0 \tau) = K\pi f_0 \tau$

So $K\pi f_0 \tau = 0$

$\Rightarrow K = 0$

We got at K=0, $|C_K| = A\tau$ .So this the maximum value of $|C_K|$

Now from the expression $K = r(T_0 / \tau)$,we get the $|C_K|$ curve will have its first zeros both negative and positive side of r = +1 and r = -1,second zeros at r = +2 and r = -2 and so on.Let consider the mid point between first and second zeros the point is

$K = \pm \left(\dfrac{T_0 / \tau + 2T_0 / \tau}{2} \right) = \pm \dfrac{3}{2} \left(\dfrac{T_0}{\tau} \right)$

At this point,

$|C_K| = A\tau \sin c \left(\pm \dfrac{3}{2}\pi \right)$

$= -\dfrac{2A\tau}{3\pi}$

both in the positive and negative side.

Next consider the points between second and third zeros in both sides.The point is

$K = \pm \left(\dfrac{2T_0 / \tau + 3T_0 / \tau}{2} \right) = \pm \dfrac{5}{2} \left(\dfrac{T_0}{\tau} \right)$

At this point,

$$|C_K| = A\tau \sin c\left(\pm\frac{5}{2}\pi\right)$$

$$=\frac{2A\tau}{5\pi}$$

both in the positive and negative side.

So obviously keeping these process next mid point will have a value equal to $-\dfrac{2A\tau}{7\pi}$ in both

sides and next one have value equal to $-\dfrac{2A\tau}{9\pi}$ in both sides and so on.

So in the conclusion we can say value of $|C_k|$ in the consecutive mid point between ZEROS will alternate in decreasing sequence .The sequence in both side

$$\text{Is} = 1/T_0\left\{-\frac{2A\tau}{3\pi},\frac{2A\tau}{5\pi},-\frac{2A\tau}{7\pi},\frac{2A\tau}{9\pi}-------\right\}$$

Another last point we want to examine the point between k=0 and first zero in both side. The

point is $k=\pm T_0/2\tau$ and $|C_k|\big|_{k=\pm T_0/2\tau} = A\tau \sin c(\pm\frac{\pi}{2})$

$$= 2A\tau/\pi \text{ in both side}$$

Considering all above facts we can plot the frequency spectrum as shown below:

Considering $\tau/T_0 = 1/2$

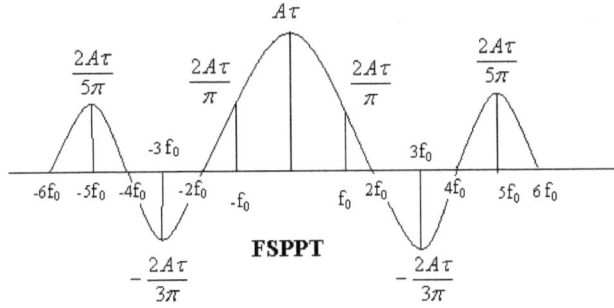

FSPPT

Let us examine the graph more closely as $\tau/T_0 = 1/2$ then k=2r.so first zero occurs at r=±1 i.e. at k=±2.Sceond zero occurs at r=±2 i.e. at k=±4 and so on .Look at the amplitude between zeros. Let us increase the time period, Let now $\tau/T_0 = 1/8$

So k=8r

So in frequency spectrum of that PPT first zero will be

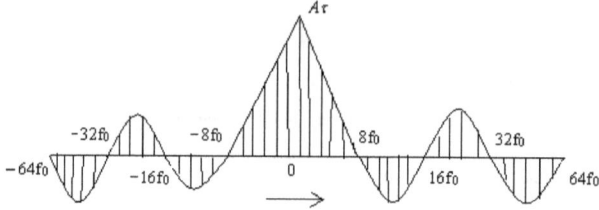

at k=± 8 for r=± 1.second zero will be at k=± 16 for r=± 2 and so on .Observe that the frequency spectrum is now more closely pact.

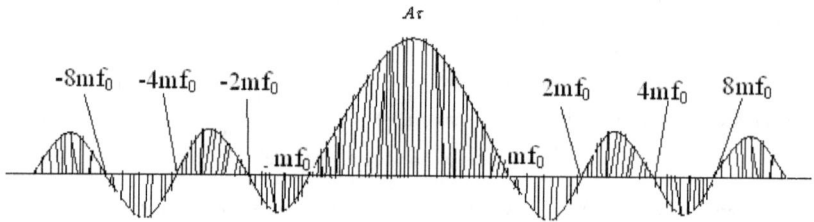

Lett us consider $\frac{\tau}{T_0} = \frac{1}{m}$ where m is a very very large integer then the frequency spectra will be almost continuous. Now keeping this fashion if $T_0 \to \infty$ i.e. $f_0 \to 0$

Then the frequency spectra will be more and more continuous and we can obtain the following ENVELOPE of continuous frequency spectra

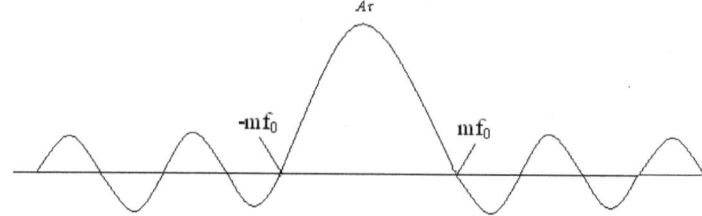

CONTINUOUS SPECTRA AND FOURIER TRANSFORM.

So in previous section we finally get the famous FSPPT. We also examine the case when time period is growing at the end we obtain the CONTINUOUS SPECTRA. letting time period equal to very very large or tends to infinity .now we shall create the Fourier series expansion of continuous spectra.

We know $X(t) = \sum\limits_{k=-\infty}^{\infty} C_k e^{jk\omega_0 t}$ ------------(A)

$$F_k = C_k / T_0$$

And $C_k = 1/T_0 \int\limits_{T_0} X(t) e^{-jk\omega_0 t} dt$ ------------------------ (A')

Obviously T_0 is the time period of signal $X(t)$ .Now let $T_0 \to \infty$ (as in continuous signal)

So $X(t) = \lim\limits_{T_0 \to \infty} \sum\limits_{k=-\infty}^{\infty} C_k e^{jk\omega_0 t}$ ------------------(B)

And $C_k = \lim\limits_{T_0 \to \infty} 1/T_0 \int\limits_{T_0} X(t) e^{-jk\omega_0 t} dt$ ---------------- (B')

Now as $T_0 \to \infty$, $\omega_0 \to 0$,then $K\omega_0 = \omega$ a new variable---------

$C_k = \lim\limits_{T_0 \to \infty} 1/T_0 \int\limits_{T_0} X(t) e^{-j\omega t} dt$

So $X(t) = \lim\limits_{T_0 \to \infty} \sum\limits_{k=-\infty}^{\infty} C_k e^{j\omega t}$

----

now we know $F_k = C_k/N_0$, substituting in (c) &(c') we get

$X(t) = \lim\limits_{T_0 \to \infty} 1/T_0 \sum\limits_{k=-\infty}^{\infty} C_k e^{j\omega t}$ ---------------(D)

And

$C_k = \lim\limits_{T_0 \to \infty} 1/T_0 \int\limits_{T_0} X(t) e^{-j\omega t} dt$ ---------------------- (D)

Now in Fourier series $\dfrac{1}{T_0} = \Delta f$ because successive harmonics are $1/T_0$ width apart.So the change

in frequency i.e $\Delta f = \dfrac{1}{T_0}$ .Now for continuous spectra as $T_0 \to \infty, \Delta f \to 0$

So

$$x(t) = \lim\limits_{\Delta f \to 0} \Delta f \sum\limits_{K=-\infty}^{+\infty} C_K e^{+j\omega t} \to E$$

$$\& \, C_K = \int\limits_{-\infty}^{+\infty} x(t) e^{-j\omega t} dt \to E'$$

Recognize the expression E is an integration with the range over $-\infty \to +\infty$ and equating $C_K = X(f)$ we finally get-

Analog Communication Engineering Fundamentals

$$x(t) = \int_{-\infty}^{+\infty} x(f)e^{+j2\pi ft} df \rightarrow F$$

$$\& \ X(f) = \int_{-\infty}^{+\infty} X(t)e^{-j2\pi ft} dt \rightarrow F'$$

This is Fourier series expression for continuous spectra(f) is called FOURIER TRANSFORM of X(f) is called FOURIER TRANSFORM of X(t)and X(t) is the INVERSE FOURIER TRANSFORM of X(f).
   NON-PERIODIC(APERIODIC) signal & FOURIER TRANSFORM:-

   Now in time domain if we take $To \rightarrow \infty$ for a periodic signal what incident would take place? Simply the periodic signal becomes an APERIODIC signal in communication theory ,if OFF time is ten times greater than ON time . We consider the signal is aperiodic.So Fourier transform is a characteristic of APERIODIC SIGNALS.

DISCRETE TIME FOURIER TRANSFORM

   We know for discrete time Fourier series

$$X[n] = \sum_{k=\langle N_0 \rangle} F_k e^{jk\omega_0 n} \ \text{------------------ (A)}$$

$$\& \ F_k = 1/N_0 \sum_{n=\langle N_0 \rangle} X[n] e^{jk\omega_0 n} \ \text{-------- (A')}$$

   where $F_k = C_k/N_0$

Now for a PERIODIC DISCRETE   time signal let $N_0 \rightarrow \infty$ i.e. the periodic signal tends to be an APERIODIC DISCRETE SIGNAL (A periodic signal which is repeated after N_0
Samples when $N_0 \rightarrow \infty$) imposing the above criteria we have found that $K\omega_0 (=k2\pi/N_0)$ Now

equal to a new variable (as $\omega_0 \rightarrow \infty$) simply ω

$$X[n] = \lim_{N_0 \rightarrow \infty} \sum_{k=\langle N_0 \rangle} F_k e^{j\omega n}$$

$$F_k = \lim_{N_0 \rightarrow \infty} 1/N_0 \sum_{k=\langle N_0 \rangle} X[n] e^{-j\omega n}$$

Now look more closely eqn(B).Summing variable is K but there is no 'k'. The expression ,now

we replace $K\omega_0$ by ω ,obviously if k's range is 0 to N_0 then range of ω is 0 to $\omega_0 N_0$ 0r 2π

   So

$$X[n] = \lim_{N_0 \rightarrow \infty} \sum_{\omega=0}^{2\pi} F_k e^{j\omega n} \ \text{-------------------(C)}$$

And

$$F_k = 1/N_0 \sum_{k=-\infty}^{\infty} X[n] e^{-j\omega n} \ \text{----------------------(C')}$$

Now we know $F_k = C_k/N_0$

substituting in (c) &(C') we get

$$X[n] = \underset{N_0 \to \infty}{Lt} \frac{1}{N_0} \sum_{\omega=0}^{2\pi} C_{Ke}^{+j\omega n} \to D$$

$$\& C_k = \sum_{n=-\infty}^{+\infty} X[n] e^{-j\omega n} \to D'$$

Now $1/N_0$ is the spacing between tne successive harmonics of discrete spectra.So change in frequency i.e $\Delta f = \dfrac{1}{N_0}$ and obviously $\Delta f \to 0$.

Replacing C_K by X[f] we have-

$$X[n] = \underset{\Delta f \to 0}{Lt} \Delta f \sum_{\omega=0}^{2\pi} X[f] e^{+j\omega n} \to E$$

$$\& X[f] = \sum_{n=-\infty}^{+\infty} X[n] e^{-j\omega n} \to E'$$

Expressing $\omega = 2\pi f$ and obviously E is an integration with limit $\omega = 0(f = 0) \to \omega 2\pi(f = 1)$ we have

$$\boxed{\begin{array}{l} X[n] = \int_0^1 X[f] e^{+j2fn} = \dfrac{1}{2\pi} \int_{2\pi} X[f] e^{+j\omega n} d\omega \to F \\ X[f] = \sum_{n=-\infty}^{+\infty} X[n] e^{-j2\pi fn} \to F' \end{array}}$$

X[f] is called the discrete time fourier transform of discrete aperiodic signal X[n] and X[n] is the inverse fourier transform of X[f].

PROPERTIES OF FOURIER TRANSFORM

1.Linearity

$$ax(t) + bx(t) \longleftrightarrow^{f} ax(f)+bx(f)$$

1.Linearity

$$aX[n] + bY[n] \longleftrightarrow^{f} aX[f] + bY[f]$$

2.Time Shifting:

$$x(t-t_0) \longleftrightarrow^{f} e^{-j\omega t_0} X(f)$$

$$\Pr oof : f[x(t-t_0)] = \int_{-\infty}^{+\infty} x(t-t_0)e^{-j2\pi ft} dt$$

$$Let \tau = t-t_0 \ \& \ dt = d\tau$$

$$\Rightarrow f[x(t-t_0)] = \int_{-\infty}^{+\infty} x(\tau)e^{-j2\pi f(\tau+t_0)} d\tau$$

$$\Rightarrow f[x(t-t_0)] = e^{-j2\pi ft_0} \int_{-\infty}^{+\infty} x(\tau)e^{-j2\pi f\tau} d\tau$$

$$\Rightarrow f[x(t-t_0)] = e^{-j\omega t_0} X(f)$$

$$\therefore x(t-t_0) \longleftrightarrow^{f} e^{-j\omega t_0} X(f) proved$$

3.Frequency shifting

$$e^{j\omega_0 t} x(t) \longleftrightarrow^{f} X(f-f_0)$$

$$\Pr oof : f[e^{j\omega_0 t} x(t)] = \int_{-\infty}^{+\infty} x(t)e^{-j2\pi(f-f_0)t} dt$$

$$let Q = f-f_0$$

$$\Rightarrow f[e^{j\omega_0 t} x(t)] = \int_{-\infty}^{+\infty} x(t)e^{-j2\pi Qt} dt$$

$$\Rightarrow f[e^{j\omega_0 t} x(t)] = X(Q) = X(f-f_0)$$

$$\therefore e^{j\omega_0 t} x(t) \longleftrightarrow^{f} X(f-f_0) proved$$

2.Time Shifting:-

$$X[n-n_0] \longleftrightarrow^{f} e^{-j\omega n_0} X(f)$$

$$\Pr oof : f(X[n-n_0]) = \sum_{n=-\infty}^{+\infty} X[n-n_0]e^{-j2\pi fn}$$

$$let m = n-n_0, \therefore n = m+n_0$$

$$\Rightarrow f(X[n-n_0]) = \sum_{m=-\infty}^{+\infty} X[m]e^{-j2\pi f(m+n_0)}$$

$$\Rightarrow f(X[n-n_0]) = e^{-j2\pi fn_0} \sum_{m=-\infty}^{+\infty} X[m]e^{-j2\pi fm}$$

$$\Rightarrow f(X[n-n_0]) = e^{-j2\pi fn_0} X(f)$$

$$\therefore X[n-n_0] \longleftrightarrow^{f} e^{-j\omega n_0} X(f) proved$$

3.Frequency shifting

$$e^{j\omega_0 n} X[n] \longleftrightarrow^{f} X[f-f_0]$$

4.Conjugation

$$x^*(t) \xleftrightarrow{\ f\ } x^*(-f)$$

4.Conjugation

$$X^*[n] \xleftrightarrow{\ f\ } X^*[-f]$$

$$\Pr oof : -X[n] = \int_0^1 X[f] e^{+j2\pi fn} df$$

$$\Rightarrow X^*[n] = \int_0^1 X^*[f] e^{-j2\pi fn} df$$

$$Now - f = f$$

$$\Rightarrow X^*[n] = -\int_1^0 X^*[-f] e^{+j2\pi fn} df$$

$$\Rightarrow X^*[n] = \int_0^1 X^*[-f] e^{+j2\pi fn} df$$

$$\therefore X^*[n] \xleftrightarrow{\ f\ } X^*[-f] \, proved$$

5.Time Reversal

$$x(-t) \xleftrightarrow{\ f\ } X(-f)$$

$$\Pr oof : X(f) = \int_{-\infty}^{+\infty} x(t) e^{-j2\pi ft} dt$$

$$\Rightarrow X(-f) = \int_{-\infty}^{+\infty} x(t) e^{+j2\pi ft} dt$$

$$Let - t = t$$

$$\Rightarrow X(-f) = -\int_{+\infty}^{-\infty} x(-t) e^{-j2\pi ft} dt$$

$$\Rightarrow X(-f) = \int_{-\infty}^{+\infty} x(-t) e^{-j2\pi ft} dt$$

$$\therefore x(-t) \xleftrightarrow{\ f\ } X(-f) \, proved$$

5.Time reversal

$$x[-n] \xleftrightarrow{\ f\ } X[-f]$$

6.Time Scaling

6.Time scaling

Let assume

$$X_k[n] = \frac{1}{K} X[n/K] \sum_{n=<K>} e^{+jK\left(\frac{2\pi}{K}\right)n}$$

From similar discussion as in TIME SCALING PROPERTY in Fourier series
We have-

$$X_k[n] = X[n/K] \quad \text{if n = multiple of K}$$

$$= 0 \qquad \text{if n} \ne \text{multiple of K}$$

$$x(at) \xleftarrow{\ f\ } \frac{1}{|a|} X\left(\frac{f}{a}\right)$$

$Proof: f[x(at)] = \int_{-\infty}^{+\infty} x(at)e^{-j2\pi ft} dt$

$Let \tau = at \ \& \ dt = \frac{d\tau}{a}$

$\Rightarrow f[x(at)] = \frac{1}{a}\int_{-\infty}^{+\infty} x(\tau)e^{-j2\pi(f/a)} d\tau$

$\Rightarrow f[x(at)] = \frac{1}{a} X(\frac{f}{a})$

Now if $f(X_k[n]) = f(X[\frac{n}{K}])$

$$= \sum_{n=-\infty}^{+\infty} X[\frac{n}{K}]e^{-j2\pi fn}$$

Let P=n/K so $\rightarrow = \sum_{n=-\infty}^{+\infty} X[P]e^{j2\pi(Kf)P}$

$= X[Kf]$

So $X_K[n] \xleftarrow{\ f\ } X[Kf]$ if n = multiple of K
Or 0 othe

7.Convolution

$[X[n] * y[n]] \xleftarrow{\ f\ } X[f]Y[f]$
proof
we know

$[X[n] * y[n]] \xleftarrow{\ f\ } \sum_{k=-\infty}^{\infty} X[k]Y[n-k]$

or X[n]*Y[n]=

$\sum_{k=-\infty}^{\infty} \int_{0}^{1} X(f)e^{jk2\pi fk} df Y(n-k)$

$= \int_{0}^{1} X(f)df \sum_{k=-\infty}^{\infty} e^{jk2\pi fk} Y(n-k)$

$= \int_{0}^{1} X(f)dfe^{j2\pi fn} df \sum_{k=-\infty}^{\infty} e^{jk2\pi fk} Y(n-k)$

$= \int_{0}^{1} X(f)Y(f)dfe^{jk2\pi fk} df$

$= f^{-1}[X[n] * y[n]]$
$[X[n] * y[n]] \xleftarrow{\ f\ } X[f]Y[f]$

7.Convolution
$[X(t) * y(t)] \xleftarrow{\ f\ } X(f)Y(f)$

we know X(f)*y(f)= $\int_{-\infty}^{+\infty} X(\tau)y(t-\tau)d\tau$

$f[X(t) * y(t)] = \int_{-\infty}^{\infty}\int_{-\infty}^{\infty} X(\tau)y(t-\tau)e^{-jk2\pi ft} d\tau dt$

now $X(\tau) = \int_{-\infty}^{\infty} X(f)e^{jk2\pi f\tau} df$

substituting in above eqn.

$f[x(t) * y(t)] = \int_{-\infty}^{\infty}\int_{-\infty}^{\infty}\int_{-\infty}^{\infty} X(\tau)y(t-\tau)e^{-jk2\pi f(t-\tau)} df d\tau dt$

let $\theta = t - \tau$ so $d\theta = -d\tau$

$= \int_{-\infty}^{\infty}\int_{-\infty}^{\infty} X(f)\int_{-\infty}^{\infty} y\theta e^{-jk2\pi f\theta} df d\theta dt$

$= \int_{-\infty}^{\infty}\int_{-\infty}^{\infty} X(f)Y(f)df dt$

=X(f)Y(f)
Hence x(t)*y(t) $\xleftarrow{\ f\ } X(f)Y(f)$

8.MULTIPLICATION:-
$[X[n]y[n]] \xleftarrow{\ f\ } X[f']Y[f-f']df'$

$f(X(n)y[n]) = \sum_{n=-\infty}^{\infty} x[n]y[n]e^{-j2\pi fn}$

now $x[n] = \int_{0}^{1} x(f') e^{j2\pi f'n} df'$

$=$

8.MULTIPLICATION:-

$x(t)y(t) \xleftrightarrow{f} X[f] * Y[f]$

proof:-

$x(t)y(t) = \int_{-\infty}^{\infty} X(f)e^{j2\pi ft}\, df\, y(t)$

$= \int_{-\infty}^{\infty} X(f)e^{j2\pi ft}\, df \int_{-\infty}^{\infty} y(f)e^{j2\pi ft}\, df$

$= \int_{-\infty}^{\infty} \int_{-\infty}^{\infty} X(f)e^{j2\pi(f+f')t} y(f)\, df$

let $\Phi = f - f'$ and so d$f' = $ dΦ

$= \int_{-\infty}^{\infty} [\int_{-\infty}^{\infty} X(f)\, y(\Phi - f)e^{j2\pi\Phi t}\, d\Phi$

$= \int_{-\infty}^{\infty} [X(f)*y[f]]\, e^{j2\pi\Phi t}\, d\Phi$

$= f^{-1}[X(f)*y[f]]$

hence $x(t)y(t) \xleftrightarrow{f} X(f) * Y(f)$

9.Differentiation in time domain:-

$\dfrac{d}{dt}X(t) \xleftrightarrow{f} j\omega X(f)$

we know $X(t) = \int_{-\infty}^{\infty} X(f)e^{j2\pi ft}\, df$

so $\dfrac{d}{dt}X(t) = \int_{-\infty}^{\infty} X(f)\, df\dfrac{d}{dt} e^{j2\pi ft}\, df$

$= \int_{-\infty}^{\infty} (j2\pi f)X(f)e^{j2\pi ft}\, df$

$= f^{-1}[j\omega\, X(f)]$

$\therefore \dfrac{d}{dt}X(t) \xleftrightarrow{f} j\omega\, X(f)$ proved

10.Integration:-

let us assume $x(t) = \dfrac{d}{dt}\left[\int_{-\infty}^{t} x(t)dt\right]$

$\sum_{n=-\infty}^{\infty} \int_{0}^{1} X(f')e^{j2\pi f'n}\, df' \int_{-\infty}^{\infty} y[n]e^{-j2\pi f'n}\, df$

$= \int_{0}^{1} X(f')df' \sum_{n=-\infty}^{\infty} y[n]e^{-j2\pi(f-f')n}$

$= \int_{0}^{1} X(f')\, y(f - f')df'$

hence X[n]Y[n] hence X[n]Y[n]

$\xleftrightarrow{f} \int_{0}^{1} X(f')\, y(f - f')df'$

9Differencing in time:-

$X[n]-X[n-1] \xleftrightarrow{f} (1-e^{-j\omega})X[f]$

Proof: -

Now $(X[n-1]) = \sum_{n=\infty}^{\infty} (X[n-1])\, e^{j2\pi fn}$

Let P=n-1 so n=P+1

$= \sum_{P=\infty}^{\infty} X[P]e^{j2\pi fP} .e^{j2\pi f}$

$= e^{j\omega} X[f]$

so

$f(X[n]-X[n-1]) = f X[n]- f X[n-1]$

$= X(f)- e^{j\omega} X(f)$

$= (1-e^{-j\omega})X(f)$

$\therefore X[n]-X[n-1] \xleftrightarrow{f} (1-e^{-j\omega})X[f]$

so $f[x(t)]=f[\frac{d}{dt}[\int_{-\infty}^{t}x(t)dt]]=j\omega\ f[\int_{-\infty}^{t}x(t)dt]$

or $x(f)=j\omega\ f[\int_{-\infty}^{t}x(t)dt]$ assuming $x(0)=0$

$\therefore\ [\int_{-\infty}^{t}x(t)dt] \xleftarrow{\ f\ } 1/j\omega\ X(f)$

11.Differentiation in frequency domain:-

$tX(t) \xleftarrow{\ f\ } j\frac{d}{d\omega}\ X(f)$

proof:

we know $X(f)=\int_{-\infty}^{\infty}\ X(t)\ e^{-j2\pi ft}\ dt$

Or $\frac{d}{dt}X(f)=-j2\pi\ tX(f)$

Or $j/2\pi\ \frac{d}{df}X(f)=t\ X(f)$

Or $f^{-1}(j\frac{d}{d\omega}\ X(f))=t\ f^{-1}\ X(f)=tX(t)$

$\therefore\ tX(t) \xleftarrow{\ f\ } j\frac{d}{d\omega}\ X(f)$

11Differentiation in frequency domain:-

$nX[n] \xleftarrow{\ f\ } j\frac{d}{d\omega}\ X[f]$

proof:- we know

$X[f]=\sum_{n=\infty}^{\infty}\ X[n]\ e^{-j2\pi fn}$

So $\frac{d}{df}X(f)=-j2\pi\ nX[f]$

So $j\frac{d}{d\omega}\ X[f]=nX[f]$

Now $f^{-1}(j\frac{d}{d\omega}\ X[f])=n\ f^{-1}\ X[f]$

Or $f^{-1}(j\frac{d}{d\omega}\ X[f])=nX[n]$

$\therefore\ nX[n] \xleftarrow{\ f\ } j\frac{d}{d\omega}\ X[f]$

MECHANICS OF FOURIER ANALYSIS

At this point we know how to analyze a signal (Periodic or aperiodic) and how to obtain its spectrum by means of a powerful toll called Fourier Analysis. Before proceeding towards the development of Fourier Analysis we have to realize by our heart the practical implementation of Fourier Analysis i.e. why and how components of a signal and the probable machinery for that purpose and also the axis of Fourier Analysis .Here GOES:-

WHY & HOW

Recall our famous form of FOURIE SERIES:-

$$x(t) = a_0 + \sum_{k=1}^{\infty} a_k \cos k\omega_0 t + b_k \sin k\omega_0 t$$

where $a_0 = 1/T_0 \int_{-T_0/2}^{T_0/2} x(t)dt = $ D.C. component=Average value of the signal over one cycle.

$$a_k = 2/T_0 \int_{-T_0/2}^{T_0/2} x(t) \cos k\omega_0 t \, dt \qquad \& \qquad b_k = \int_{-T_0/2}^{T_0/2} x(t) \sin k\omega_0 t \, dt$$

let for example

$$x(t) = a_m \cos m\omega_0 t + b_m \sin m\omega_0 t \, dt + a_n \cos n\omega_0 t + b_n \sin n\omega_0 t + a_p \cos p\omega_0 t + b_p \sin p\omega_0 t$$

(where m,n& p are integers.)

again recall the following relations:-

1. $\int_{-T_0/2}^{+T_0/2} \cos(m\omega_0 t)\cos(k\omega_0 t)dt = [T_0/2 \qquad m=k$

$[\, 0 \qquad m \neq k$

2. $\int_{-T_0/2}^{+T_0/2} \cos(m\omega_0 t)\sin(k\omega_0 t)dt = 0$

3. $\int_{-T_0/2}^{+T_0/2} \sin(m\omega_0 t)\sin(k\omega_0 t)dt = [\, T_0/2 \qquad m=k$

$[0 \qquad m \neq k$

Now $a_k = 2/T_0 \left[\int_{-T_0/2}^{T_0/2} \{\cos(m\alpha\omega_0 t) + a_n \cos(n\omega_0 t) + a_p \cos(p\omega_0 t)\}\cos k\omega_0 t \, dt + \int_{-T_0/2}^{T_0/2} \{\sin(m\omega_0 t) + b_n \sin(n\omega_0 t) + b_p \sin(p\omega_0 t)\}\sin k\omega_0 t \, dt \right]$

Now from the relation 2. We can readily say underlined portion is equal to 0

So

$a_k = 2/T_0$

$$\int_{-T_0/2}^{T_0/2} a_m \cos(m\omega_0 t)\cos(k\omega_0 t)dt + \frac{2}{T_0} a_n \cos(n\omega_0 t)\cos(k\omega_0 t)dt + \frac{2}{T_0}\int_{-T_0/2}^{T_0/2} a_p \cos(p\omega_0 t)\cos(k\omega_0 t)dt$$

Now if we set k=m then last two integration will vanish leaving $a_k=a_m$ i.e. by setting k=m (or n or p) ,we can find a_m (or $a_n$or a_p)i.e. the $n\omega_0$ frequency component (or $n\omega_0$ or $p\omega_0$).physically we can see that for m=k the first integration leads

$$a_k = 2/T_0\, a_m \int_{-T_0/2}^{T_0/2} \cos^2 k\,\omega_0 t\, dt \quad \text{(other two integration vanished)}$$

now we know unlike -1 $<\cos$ m ω_0t <1, 0< $\cos^2$m $\omega_0 t$ <1

Hence $\cos^2$m $\omega_0 t$ has a non-zero average   value or D.C. value now and (0< $a_m \cos^2$m $\omega_0 t < a_m$)

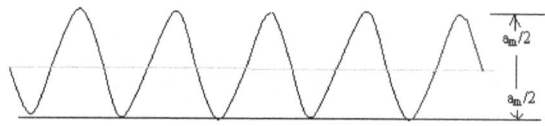

$$a_m \cos^2 m\,\omega_0 t = a_m/2 \cos 2m\,\omega_0 t + a_m/2$$

from the graph $a_m \cos^2 m\,\omega_0 t$

we can see that it is a combination of a sinusoid $a_m/2 \cos 2m\,\omega_0$t and a D.C. level of $a_m/2$. now if

we integrate $a_m \cos^2 m\,\omega_0 t$ over a time period then $\int_{-T_0/2}^{T_0/2} a_m \cos^2 m\,\omega_0 t\, dt = \int_{-T_0/2}^{T_0/2} a_m/2$

$$\cos^2 m\,\omega_0 t\, dt + \int_{-T_0/2}^{T_0/2} a_m/2 dt$$

$$= 0 + a_m/2.T_0 = a_m/2.T_0$$

Multiplying the above integration by $2/T_0$(as in a_k) .we get a_m which is the amplitude of $m\omega_0$ frequency component. Hence SQUARING OF SINUSOID PRODUCE A D.C.LEVEL which is proportional to related frequency components.

Now take the   physics of vanishing,Let k=m,hence the 2nd integration of a_k is

$$\frac{2}{T_0} \int_{-T_0/2}^{T_0/2} \cos(n\omega_0 t)\cos(m\omega_0 t)dt$$

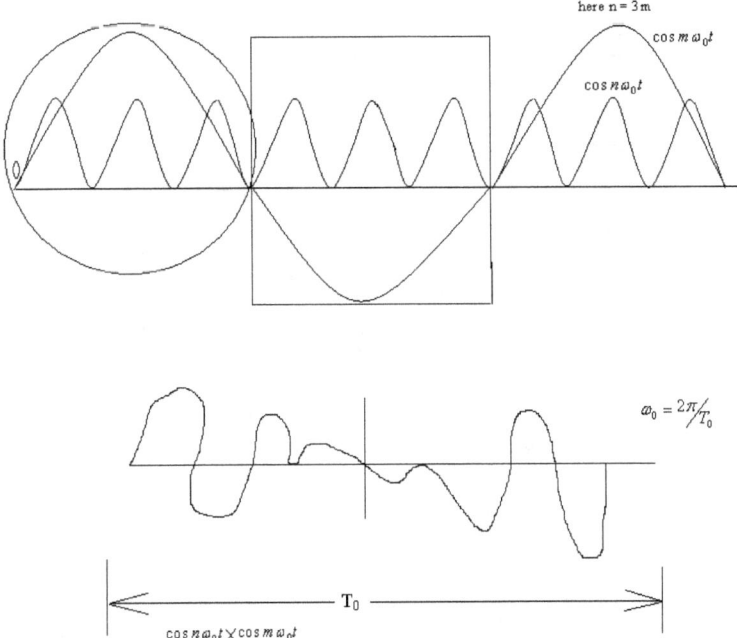

In the first diagram we have shown the $\cos m\omega_0 t$ and $\cos n\omega_0 t$ and in the 2nd diagram we have shown $\cos m\omega_0 t \cos n\omega_0 t$ diagram (hypothetical, not actual)we can see from the 2nd diagram the first half(0 to $T_0/2$)is just equal and opposite of 2nd half($T_0/2$ to T_0)Hence if we integrate the second graph over T_0 then it results 0 ,because it has 0 D.C. level or it is simply symmetric about the horizontal axis or simply has 0 average value etc. Hence if we set k=m then we can get only a_m not $a_n$or a_p or anything, this is the mechanics of Fourier analysis. Hence from the above conclusion we can design a hypothetical FOURIER ANALYZER as follows:-

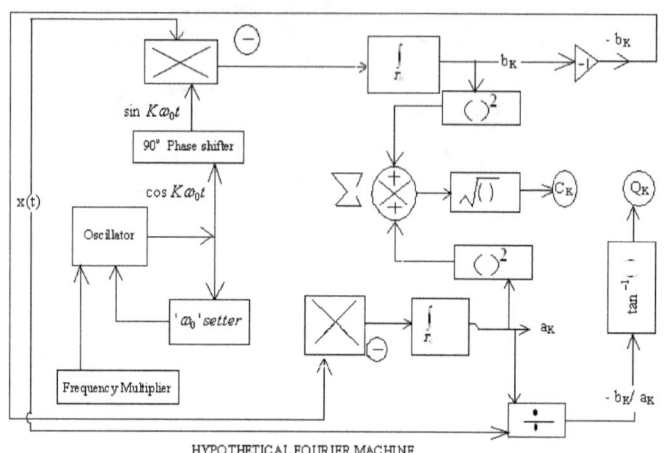

HYPOTHETICAL FOURIER MACHINE

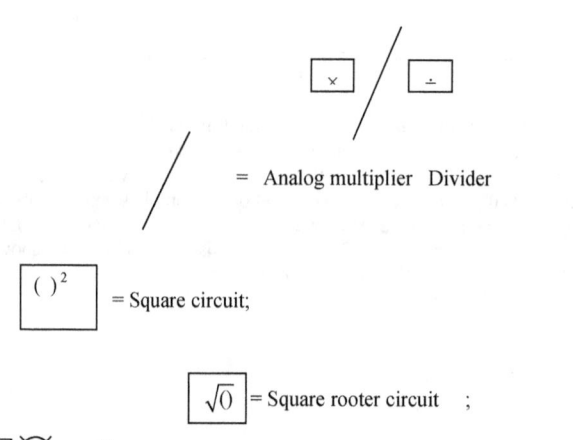

= Analog multiplier   Divider

$( )^2$ = Square circuit;

$\sqrt{0}$ = Square rooter circuit   ;

$\sum \bigotimes$ = adder;

Analog Communication Engineering Fundamentals

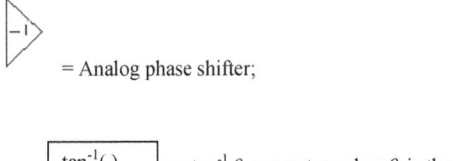

= Analog phase shifter;

$\boxed{\tan^{-1}(\ )}$ $= \tan^{-1}\theta$ generator, when θ is the input

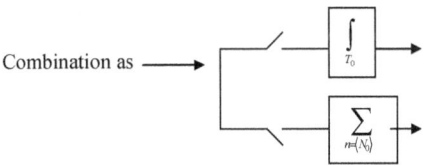
= Analog integrator

a digitally controlled switch ⊘ remain closed up to T_0 duration then

opened(MOSFETswitch).For Discrete time signal(i.e. discrete Fourier series)we just have to replace $\boxed{\int_{T_0}}$ by $\boxed{\sum_{n=\langle N_0\rangle}}$ we can use automatic switching

Combination as ⟶ [switch — $\int_{T_0}$ →] [switch — $\sum_{n=\langle N_0\rangle}$ →]

(when upper switch is open lower switch
have to be closed and vice-versa)
when x(t) is an aperiodic signal above machine automatically perform FOURIER TRANSFORM.

At this point now we can easily realize that how Fourier analysis works practically to analyze a signal's spectrum for a periodic signal x(t).We first have to calculate the fundamental period manually ,then setting the oscillator at that fundamental frequency ω_0 then varying the k=

Analog Communication Engineering Fundamentals

0,1,2,3.... We can obtain D.C. value(C_0) first harmonic or fundamental (C_1) second harmonic(C_2) and so on .Actually $k\omega_0$(k=0,1,2,3…….) frequency is obtained by using an electronic circuit known as frequency multiplier.

LTI system & FOURIER ANALYSIS

Do u know any signal(continuous or discrete time)is a composition of complex exponential time signal? Let us define two constants as follow s= $s=\sigma + j\omega$ and $z=Ae^{j\omega}$

Where σ is called 'damping co-efficient'(a constant, real) ω is called angular frequency $|z| = A$ =constant 's' is called complex frequency in continuous time and z may be called a complex exponential. Similarly $e^s = e^{\sigma + j\omega}$ is a complex exponential

Let x(t)= e^{st};if x(t) is the input of an LTI system what will be the y(t)? Let us calculate it----- we know for a LTI system y(t)=x(t)*h(t)

Or y(t)=
$$\int_{-\infty}^{\infty} h(\tau)e^{s(t-\tau)}d\tau$$
$$e^{st}\int_{-\infty}^{\infty} h(\tau)e^{-s\tau}d\tau$$

$x(n) = z^n$
$$y(n) = \sum_{k=-\infty}^{\infty} h(k)z^{(n-k)}$$
$$= z^n \sum_{k=-\infty}^{\infty} h(k)z^{-k}$$

Now one can say $\int_{-\infty}^{\infty} h(\tau)e^{-s\tau}d\tau$ | $\sum_{k=-\infty}^{\infty} h(k)z^{-k}$ is independent of time t/n and may be regarded as a

constant with respect to time .Let us assume H(s)= $\int_{-\infty}^{\infty} h(\tau)e^{-s\tau}d\tau$ $|H(z) = \sum_{k=-\infty}^{\infty} h(k)z^{-k}$,so we have

y(t)=h(s)e^{st} and y(n)=H(z)z^n

Now for Fourier Transform S=$j\omega$ and z=$e^{j\omega}$ (i.e. $\sigma = 0 and |z| = A = 1$) hence we can say that for an LTI system—

Y(t)=H(jω) $e^{j\omega t}$ and y(n)= H(jω) $e^{j\omega n}$
[Here H(jω)$\leftrightarrow$h(t) and H($e^{j\omega}$) $\leftrightarrow$ h[n]]
for x(t) = $e^{j\omega t}$ and x[n] = $e^{j\omega n}$
Let x(t) = $C_1(e^{j\omega_1 t} + e^{-j\omega_1 t}) + C_2(e^{j\omega_2 t} + e^{-j\omega_2 t}) + C_3(e^{j\omega_3 t} + e^{-j\omega_3 t})$

Similarly x[n]= $C_1(e^{j\omega_1 n} + e^{-j\omega_1 n}) + C_2(e^{j\omega_2 n} + e^{-j\omega_2 n}) + C_3(e^{j\omega_3 n} + e^{-j\omega_3 n})$

Now as X(jω) = $\int_{-\infty}^{+\infty} x(t)e^{-j\omega t}dt$ / X($e^{j\omega}$) = $\sum_{n=-\infty}^{+\infty} x[n]e^{-j\omega n}$

By setting $\omega = \omega_1, \omega_2, \omega_3$-----------one by one we can obtain in the same way C_1, C_2, C_3…..
.This is FOURIER TRANSFORM mechanics logically same to FOURIER SERIES mechanics described earlier.

AXIS OF FOURIER ANALYSIS

Basically we know both for Fourier Series and Fourier Transform we vary the k(for Fourier series)or vary the ω simply(for Fourier transform).now at this point we know about significance and meaning of REAL FREQUENCY (f as $\cos\omega$ t),exponential or rotating frequency($j\omega$ .as $e^{j\omega t}$) and complex frequency(s,s=$\sigma + j\omega$ as e^{st}).We also know how equal and opposite exponential can be combined and produce a real frequency of amplitude twice as of the exponential frequency(as $\dfrac{A}{2}e^{-j\omega_0 t} + \dfrac{A}{2}e^{j\omega_0 t} = A\cos\omega_0 t$)

Now Fourier series(or better u are a developed person now ,it is Exponential Fourier Series) and Fourier Transform both analyze the spectrum on $j\omega$ axis (remember $e^{jk\omega_0 t}$ or $e^{jk\omega_0 n}$ & $e^{j\omega t}$ or $e^{j\omega n}$ for Fourier series and transform respectively) because Fourier analysis is a tool which help us to find the Exponential frequency components in a periodic or aperiodic signal Conventionally imaginary axis is drawn vertically and real axis is drawn horizontally, but for Fourier analysis there is no real frequency component (unlike s=$\sigma + j\omega$ complex frequency),hence we can draw $j\omega$

axis horizontally by omitting 'j' simply write 'ω' or more simply for Fourier series 'k'(represent kω_0) as a variable, but you should always remember this is originally '$j\omega$' axis.

D.C. component and FOURIER analysis:-
We know for a periodic signal x(t) or x(n)----

$$C_k = \frac{1}{T_0}\int_{T_0} x(t)e^{-jk\omega_0 t}\,dt \quad \text{or} \quad C_k = \frac{1}{N_0}\sum_{n=\langle N_0\rangle} x(n)e^{-jk\omega_0 n}$$

$$\text{now } C_0 = \frac{1}{T_0}\int_{T_0} x(t)\,dt \quad \text{or} \quad C_0 = \frac{1}{N_0}\sum_{n=\langle N_0\rangle} x(n)$$

We know for a periodic signal D.C. component is simply the net area over1-time period, which

is simply depicted in equation of C_0 .this net area actually signifies total charges transported

through 1-time period. Also we can see the D.C. component has the unit of voltages or current

Students always lead to a chaos when they try to find the D.C. component of larger

class of periodic signal by using 'C_k' relations directly. all they have at last is some wrong

answers and more bad conception.

It is due to the fact that for larger classes of signal after performing the integration (as for

triangular wave ,asymmetric square wave and so on for 90% cases) 'k' often comes out at the

denominator ,hence k=0 leads to an hoax state, another reason is that there is no linear

Analog Communication Engineering Fundamentals

relationship between x(t)and $x(t)e^{-jk\omega_0 t}$,hence integration of this two over a time period and substitution of value leads two two different result contrast to substitution before integration. So it is always recommended to the students that they should find the D.C. component or A_0(or a_0)

first by the simple formulae of its own i.e. $A_0 = \frac{1}{T_0} \int_{-T_0/2}^{T_0/2} x(t)dt$ or $A0 = \frac{1}{N_0} \sum_{n=\langle N_0 \rangle} x(n)$ f or a

periodic signal then they can find C_k(orA_k,a_k) by using the usual exponential series formula of

$$C_k = \frac{1}{T_0} \int_{-T_0/2}^{T_0/2} x(t)e^{-jk\omega_0 t} dt \quad \text{or} \quad C_k = \frac{1}{N_0} \sum_{n=\langle N_0 \rangle} x(n)e^{-jk\omega_0 n}$$

Chapter 2

DETERMINISTIC SIGNAL ANALYSIS and AN INTRODUCTION TO RANDOM SIGNAL THEORY

Study of Random Signal Theory is divided in three parts –

      1.   PROBABLITY THEORY
      2.   SIGNAL ALALYSIS
      3.   RANDOM PROCESSES

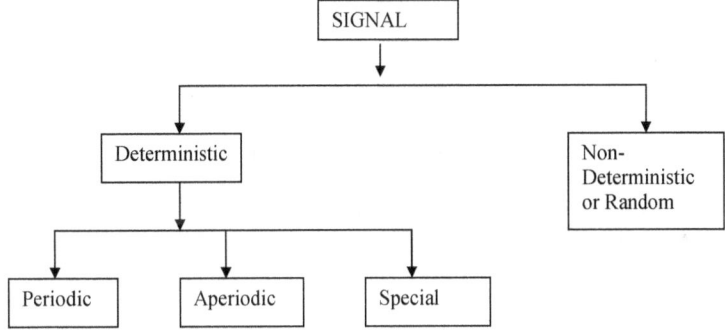

Non – Deterministic property of Random Signal :-

        A Signal can be said RANDOM if its character (i.e. amplitude, power or energy, duty cycle, pulse width etc. but mostly we can rely on amplitude only) can't be determined uniquely on time. This means that – we cannot predict at which time the Variable takes which Value – But it should be clearly under stood that the Range of Values which such a variable can take should be specified. For example – A dice can takes only 6 – different values i.e. 1 2 3 4 5 & 6, but we don't know when it will take. Similarly take the case of Line Voltage. We know normally it is within the range of 190 – 230 volts. But we can't say what will be the actual value

Analog Communication Engineering Fundamentals

of line voltage at any time. The first example it for DISCRETE RANDOM VARIABLES (DRV) and second is for CONTINUOUS RANDOM VARIABLES (CRV) .We will learn more about DRVs and CRVs at latter.

For Random Signal we again want to know the Information such as Energy, Power, Correlation, etc. But there is a Problem. For Periodic or Aperiodic Signal we know the time characteristics of signal and from those characteristics we can easily obtain information about average power, Total Energy, correlation etc. But for the Random signal time characteristics has no meaning so we have to take the help of probability.

Take again the case of a DICE. Let we have to calculate average value of the DICE. Number. In 'm' no of throwing if m = 6 then we can say the average value of Dice say

$$n_{av} = (n_1 + n_2 + n_3 + \text{-----} + n_m) / m = (n_1 + n_2 + n_3 + n_4 + n_5 + n_6)/6$$

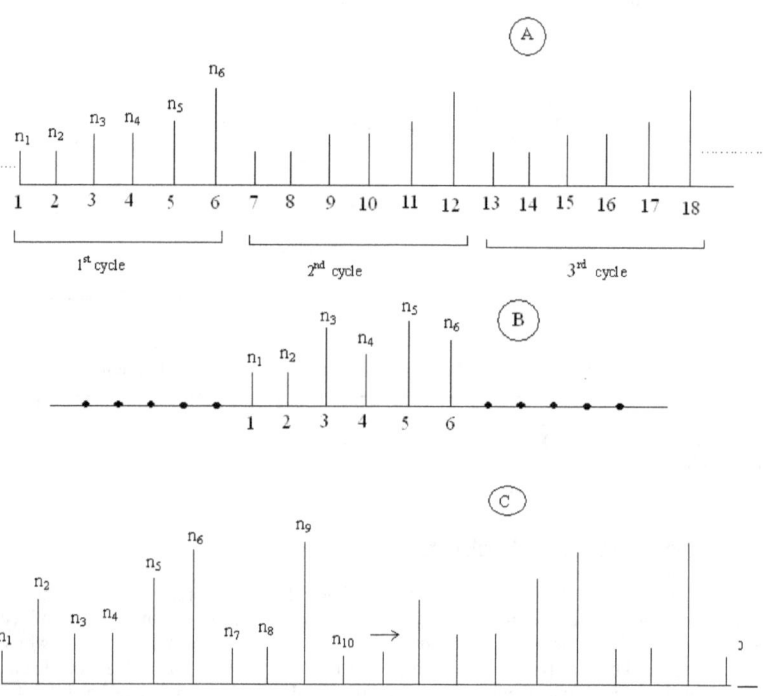

Analog Communication Engineering Fundamentals

A) Hypothetical Repeated (periodic) sequence of a Dice.
B) Practical aperiodic sequence of a Dice.
C) Practical RANDOM Sequence of a Dice.

Now here we consider 3 – General Cases. A) It's a hypothetical model. Here we assume that Dice no. will occur in a periodic sequence of a sequence period of 6. Hence the average value of N will be

$$n_{av} = \lim_{k \to \infty} (n_1 \times k + n_2 \times k + n_3 \times k + n_4 \times k + n_5 \times k + n_6 \times k)/6 \times k$$

$$(k = \text{no. of cycle})$$
$$= (n_1 + n_2 + n_3 + n_4 + n_5 + n_6)/6$$

Hence without <u>WAITING</u> for all the cycle indefinitely we can <u>easily</u> calculate the average value

of the number from the above formula. It is clear that this is the case of periodic wave form

where from the property of a cycle we can calculate the property of total waveform. B) It is a

practical model. Here we taken the case that the Dice is thrown a finite no of times (say 6 times

here). Hence we have to wait up to the finishing point of the sequence and then perform the

averaging operation according to the $n_{av} = (n_1 + n_2 + n_3 + n_4 + n_5 + n_6)/6$. The difference between

A) & B) is in B) we have to wait for the end of the sequence unlike of A). But in B) the sequence

is finite, hence we doesn't have to wait for an indefinite or a very long time. Hence we can easily

determine the average value after receiving all sequence clearly. This is the case of an

APERIODIC WAVE FORM. At this very important point of Junction we must know what is the

basic difference between an APERIODIC signal and a RANDOM signal. From the view of

signal processing I want to say that the aperiodic waves also have two different categories like

Periodic wave -

PERIODIC WAVE

| WAVE with a mathematically known or analytical shape. | WAVE with an irregular shape. |

APERIODIC WAVE

| WAVE with a pulse shape, which is mathematically recognizable. | WAVE with an irregular pulse shape |

Those 4 categories of waveforms are shown in figure.

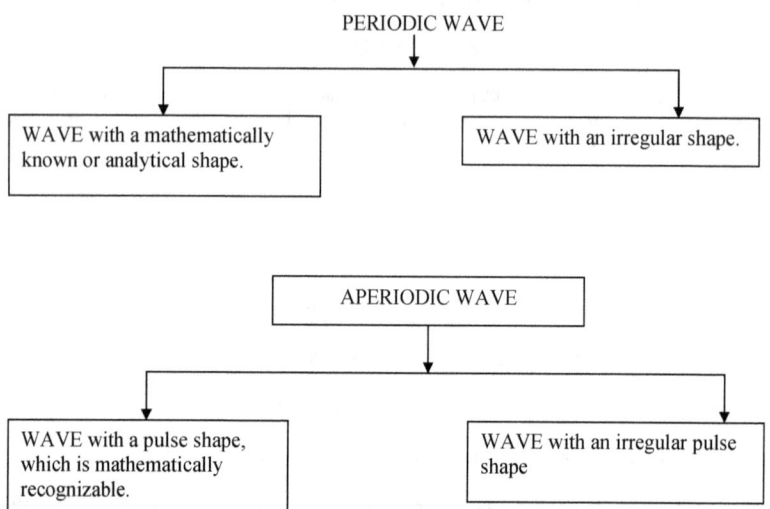

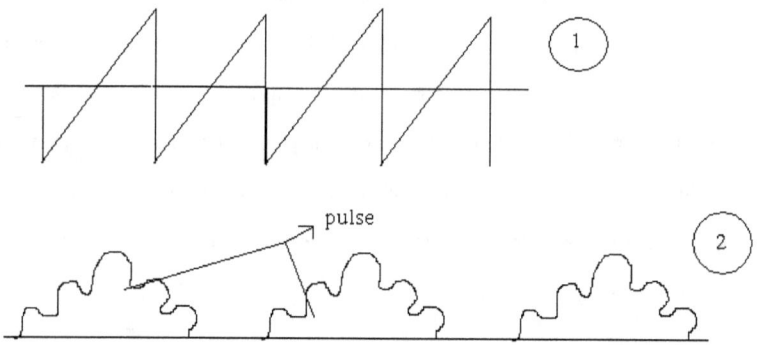

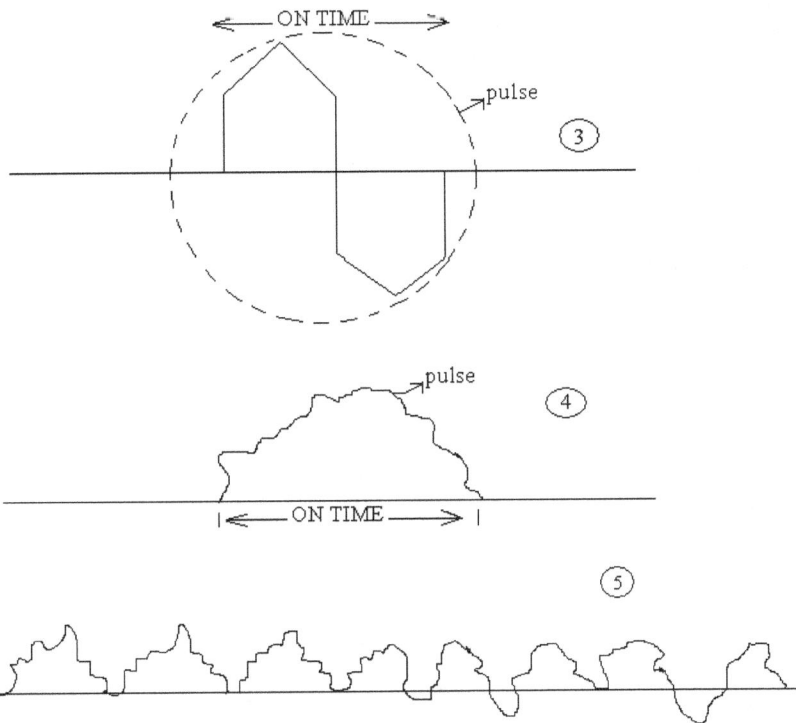

Now if for a periodic waveform the ratio 0 < PULSE ON TIME / PULSE OFF TIME < 0.1 then the wave in signal processing considered as an APERIODIC WAVE. Now the wave form of 4th category is an Aperiodic Wave, if the PULSE of this wave form exist for a very long time i. e. violets the above ratio and also if the next pulse is not of the same nature as of the previous one and again this pulse sequence continued for a very considerable indefinitely long time (see wave form of 5). Hence we can view a RANDOM SIGNAL (Wave form of 5) as an APERIODIC wave of very very long pulse duration of an unknown or irregular shape or can be considered as a conglomeration of various types of APERIODIC waves of different pulse shape. Hence Random types of signal neither have periodic nature nor have finite duration like aperiodic wave, even it normally do not have regular recognizable shape, i.e. we can not say exactly what will be the value of signal at a specific time from the past.

But for any signal whether its deterministic (periodic or Aperiodic) or non – deterministic (Random) we need to find some parameter related to that signal like → energy or power, mean value, RMS Value, Convolution, Correlation etc. without finding the above parameter. One cannot process any signal. Techniques for finding out the P_{av}, E_T , x (t) * y (t), R_{xy} (τ) d.c. Value, for the deterministic signal is already discussed by some how the 'integration' process over time axis technique but one can not apply the some technique for Random signal because it neither has periodic nature nor they have any recognizable shape so what can we do for them ? Before proceeding to the solutions for Random signal we just take another look at the deterministic signal and just see how we could find the above-mentioned things for them.

Analysis of Deterministic Signal.

Let X (t) is the deterministic Signal. If it is periodic it has a time period of To (No) for both periodic and Aperiodic cases we gave assumed the signal has total time span $\left(\underset{T \to \infty}{Lt} \right) T$ or $\left(\underset{N \to \infty}{Lt} \right) N$

PERIODIC SIGNAL (POWER SIGNAL)	APERIODIC SIGNAL (ENERGY SIGNAL)
1. MEAN VALUE(D.C value) $\Rightarrow 1/T_0 \int_{-T_0/2}^{+T_0/2} x(t)\,dt$ $\Rightarrow 1/N_0 \sum_{n<N_0>} x[n]$ (both values are finite value, otherwise it will be stated)	**1. MEAN VALUE (D.C. Value)** $\Rightarrow \left(\begin{array}{c} Lt \\ T \to \infty \end{array}\right) 1/T \int_{-T_0/2}^{+T_0/2} x(t)\,dt = 0$ $\Rightarrow \left(\begin{array}{c} Lt \\ N \to \infty \end{array}\right) 1/N \sum_{n<N_0>} x[n] = 0$ (because aperiodic signal has finite pulse duration, let us assume τ is the pulse duration)
2. RMS Value $\Rightarrow \sqrt{\ } \; 1/T_0 \int_{-T_0/2}^{+T_0/2} x^2(t)\,dt$ $\Rightarrow \sqrt{\ } \; 1/N_0 \sum_{n<N_0>} x^2[n]$	**2. RMS Value** $\Rightarrow \sqrt{\ } \left(\begin{array}{c} Lt \\ T \to \infty \end{array}\right) 1/T \int_{-T_0/2}^{+T_0/2} x^2(t)\,dt = 0$ $\Rightarrow \sqrt{\ } \left(\begin{array}{c} Lt \\ N \to \infty \end{array}\right) 1/N \sum_{n<N_0>} x^2[n] = 0$
3. TOTAL ENERGY $E_T = \begin{array}{c} Lt \\ T \to \infty \end{array} \int_{-T/2}^{+T/2} x^2(t)\,dt = \infty$ $E_T = \begin{array}{c} Lt \\ N \to \infty \end{array} \sum_{n<N>} x^2[n] = \infty$	**3. TOTAL ENERGY** $E_T = \begin{array}{c} Lt \\ T \to \infty \end{array} \int_{-T/2}^{+T/2} x^2(t)\,dt = \int_{-\tau/2}^{\tau/2} x^2(t)\,dt$ $E_T = \begin{array}{c} Lt \\ N \to \infty \end{array} \sum_{n<N>} x^2[n] = \sum_{n<\tau>} x^2[n]$
4. AVERAGE POWER $P_{av} = 1/T_0 \int_{-T_0/2}^{+T_0/2} x^2(t)\,dt$ $P_{av} = 1/N_0 \sum_{n<N_0>} x^2[n]$	**4. AVERAGE POWER** $P_{av} = \left(\begin{array}{c} Lt \\ T \to \infty \end{array}\right) 1/T \int_{-T_0/2}^{+T_0/2} x^2(t)\,dt = 0$ $P_{av} = \left(\begin{array}{c} Lt \\ N \to \infty \end{array}\right) 1/N \sum_{n<N_0>} x^2[n] = 0$ (Because it has finite pulse duration)
5. CONVOLUTION (Both signal should have same time period of T_0) $\Rightarrow Z(t) = x(t) * y(t)$ $= 1/T_0 \int_{-T_0/2}^{+T_0/2} x(\tau)y(t-\tau)\,d\tau$ $\Rightarrow Z[n] = 1/N_0 \sum_{k<N_0>} x(k)y(n-k)$	**5. CONVOLUTION** $\Rightarrow Z(t) = x(t) * y(t)$ $= \int_{-\infty}^{+\infty} x(\tau)y(t-\tau)\,d\tau$ $\Rightarrow Z[n] = \sum_{k=-\infty}^{+\infty} x(k)y(n-k)$
6. CORRELATION $R_{xy}(\tau) = 1/T_0 \int_{-T_0/2}^{T_0/2} x(t)y(t+\tau)\,dt$ $R_{xy}(k) = 1/N_0 \sum_{k<N_0>} x(n)y(n+k)$ (Both signal should have same time period T_0 or N_0)	**6. CORRELATION** $R_{xy}(\tau) = \int_{-\infty}^{+\infty} x(t)y(t+\iota)\,dt$ $R_{xy}(k) = \sum_{k=\infty} x(n)y(n+k)$

Now we have already studied first 5 things extensively in our previous course. The new things is CORRELATION.

CORRELATION BETWEEN WAVE FORMS

CORRELATION between two different Wave forms = CROSS – CORRELATION of simply called as CORRELATION.

CORRELATION between the same Wave = AUTOCORRELATION.

Q.)   WHAT IS CORRELATION?

Ans.)   Correlations between the waveforms mean"a measurement or indication of the relation between the waveforms". Correlation tries to find out the dependency or better say relatively between two or more physical Quantity. Let's take the variable like Father's AGE and Son's age. When first one is increased second one also increases. In this sense we can say they are positively correlated take the quantity like demand and supply. If one increases the other decrease; we can say they are negatively correlated.   But whether positive or negative there are some relations between the two. Now if we take the quantity like 'SPEED of a CAR' and 'Input Impedance of a Two PORT network' there is never ever any relation between the two. Simply we can say they are uncorrelated.

Q.)   How could we found the correlation between two signals WAVEFORM?

Ans.)   To find the correlation between waveforms x (t) and y (t) shown below---- we simply shift one waveform and the other waveform say x (t) stays as it is →

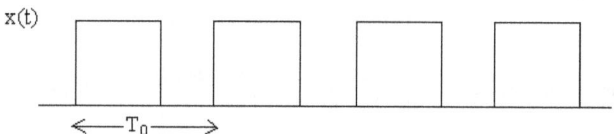

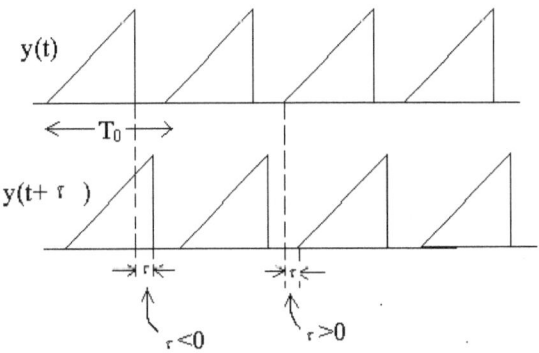

CORRERLATION BETWEEN TWO PERIODIC WAVEFORMS

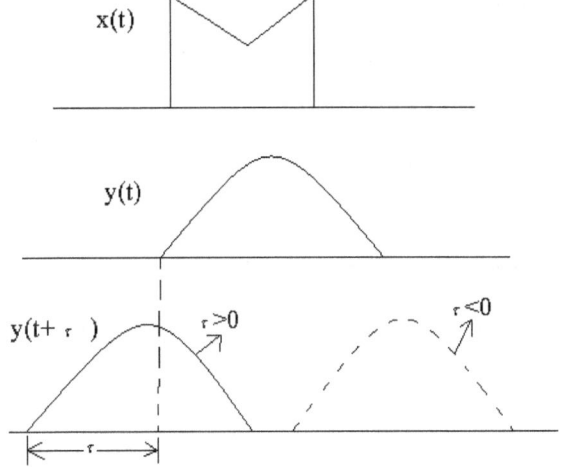

CORRELATION BETWEEN TWO APERIODIC WAVEFORMS

(It should be noted that Correlation between two periodic waves is not possible if they don't have same time period. We have to use aperiodic correlation). → Shift the other wave i.e. y (t) by τ sec. 'τ' is called 'shifting parameter'. τ may be positive or negative. Now for both periodic and aperiodic wave correlation function is defined as follows →

For periodic wave:-

$$R_{xy}(\tau) = 1/T_0 \int_{-T_0/2}^{+T_0/2} x(t)y(t+\tau)dt \qquad \text{for } \tau > 0$$

$R_{yx}(\tau) = 1/T_0 \int_{-T_0/2}^{+T_0/2} x(t)y(t+\tau)dt$ for $\tau < 0$

For aperiodic wave: -

$R_{xy}(\tau) = \int_{-\infty}^{+\infty} x(t)y(t+\tau)dt$ for $\tau > 0$

$R_{yx}(\tau) = \int_{-\infty}^{+\infty} x(t)y(t+\tau)dt$ for $\tau < 0$

Now for simplicity we ignore the scaling factor 1/To. Both the equations take same form. Widely we can say that this is the time or better say DETERMINISTIC 'mean' value of x (t) y (t + τ)
i.e. $\overline{x(t)\,y(t+\tau)}$ [note that time or deterministic 'mean' value of some quantity denotes as (.)]. Hence regard less of value of τ →

$R_{xy}(\tau) = \overline{x(t)\,y(t+\tau)}$

The above function $R_{xy}(\tau)$ is called the cross – correlation between waveforms x (t) & y (t). In general expression we can say x (t_1) y (t_2) = R_{xy} ($t_2 - t_1$)

Now in Rxy (τ) if we replace y by x i.e. $R_{xx}(\tau)$ =   <x(t)x(t+τ)>; [note symbols <> and ~ used mutually indiacting time domain averaging operation]   this is called the AUTO CORRELATION of wave form x (t).

Q.)   What is the power of two periodic signal $[V_1(t) + V_2(t+\tau)]$ of time period T_0 each?

Ans) Now $P_1 = 1/T_0 \int_{-T_0/2}^{+T_0/2} V_1^2(t)dt$; $P2 = 1/T_0 \int_{-T_0/2}^{+T_0/2} V_2^2(t)dt$

Or, $P_2 = 1/T_0 \int_{-T_0/2}^{+T_0/2} V_2^2(t+\tau)dt$

Let P = power of $[V_1(t) + V_2(t+\tau)]$

$= 1/T_0 \int_{-T_0/2}^{+T_0/2} (V_1 + V_2)^2 dt = 1/T_0 \int_{-T_0/2}^{+T_0/2} (V_1^2 + V_2^2 + 2V_1V_2)dt$

$= 1/T_0 \int_{-T_0/2}^{+T_0/2} V_1^2(t)dt + 1/T_0 \int_{-T_0/2}^{+T_0/2} V_2^2(t+\tau)dt + 2.\times 1/T_0 \int_{-T_0/2}^{+T_0/2} V_1(t)V_2(t+\tau)dt$

[Assuming τ>0]

$= P_1 + P_2 + 2\,R_{12}(\tau)$

Now for uncorrelated waveform $R_{12}(\tau) = 0$ then power of the combination of two waveform simply equals to $P = P_1 + P_2$

$R_{xx}(\tau) = <x(t)x(t+\tau)>$

Now $R_{xx}(0) = \overline{x^2(t)}$= Power or Energy of the signal x(t)

$R_{xy}(\tau) = <x(t)y(t+\tau)>$

[τ >0]

Now $R_{xy}(-\tau) = <x(t)y(t-\tau)>$

Let $t - \tau = \theta$;

$= x(\tau+\theta)y(\theta)$

$= R_{yx}(\tau)$

Hence	$R_{xy}(-\tau) = R_{yx}(\tau)$
Or	$R_{yx}(-\tau) = R_{xy}(\tau)$

But in general, $R_{xy}(\tau) \neq R_{xy}(\tau)$

$$\Rightarrow \quad R_{xx}(\tau) = \overline{x(t)x(t+\tau)}$$

$$\text{So } R_{xx}(-\tau) = \overline{x(t)x(t-\tau)}$$

Let $\theta = t - \tau$

$$= \overline{x(\theta + \tau)x(\theta)}$$

$$= \overline{x(\theta)x(\theta + \tau)}$$

$$= R_{xx}(\tau)$$

or $\boxed{R_{xx}(\tau) = R_{xx}(-\tau)}$

Hence $R_{xx}(\tau)$ is the even function of τ.
Now we can say for x(t)-

$$\overline{[x(t) - x(t+\tau)]^2} \geq 0$$

$$\overline{x^2(t) + x^2(t+\tau) - 2x(t)x(t+\tau)} \geq 0$$

$[$Now $R_{xx}(0) = \overline{x^2(t)} = \overline{x^2(t+\tau)}]$
or , $R_{xx}(0) + R_{xx}(0) - 2 R_{xx}(\tau) \geq 0$
$\therefore R_{xx}(0) \geq R_{xx}(\tau)$
$\therefore R_{xx}(0)$ is the maximum value of $R_{xx}(\tau)$
Now $R_{xx}(0)$ = power or energy of the signal

i.e. if we do not shift the signal (i.e. $\tau = 0$) then the Autocorrelation of the signal x(t) is maximum and it simply equals to the power or energy of the signal, and this is desirable ,because if we multiply any signal with itself and then evaluating the mean then it simply gives the Energy or power of the signal.

Q) What is the difference between CONVOLUTION & CORRELATION?
Ans.) At this important point we must clearly understand the difference between the convolution & correlation. Convolution and correlation basically depends on same mechanics. Shifting one signal w.r.t. Other by 'τ' then multiplying them and then integrate them over the time axis for correlation and on the 'τ' axis for convolution. Both of the 'C₀' finds the mutual relation between two signal one 't' [for convolution as Z(t)= x(t)*y (t)] and other on shifting parameter 'τ' itself [for correlation $R_{xy}('\tau')$]. Convolution shows how the relation between two signal changed on time 't' for fixed 'τᵢ' and correlation shows the changing of mutual relation at a particular time on the shifting parameter 'τ'.

ENOUGH of DETERMINISTIC –Let's talk about RANDOM

See how we can easily obtain all the required parameter for deterministic signal; but all the tools (i.e. finding 'mean' of something which is related to that parameter over time) used for deterministic signal terminating for RANDOM SIGNAL for the reason previously described i.e. because of Random Signal has no particular recognizable or determinable shape over time neither it has finite duration. So what we can do for RANDOM SIGNAL?

We turn for the Random Signal Analysis towards the PROBABILITY THEORY from a large sample of previously collected values for any randomly changing physical phenomenon we first calculate the probability of the occurrence of any particular value (for discrete random variable) & any particular range of values (for continuous Random variables) and using that probability related data. We can calculate the corresponding probability density function (PDF) for the CRV and set of all probability value for the $\overline{DRV}$ using the above we can calculate mean (d.c. value), mean square value (Power or Energy), convolution, correlation and so on. But one thing should be remember, if the sample space or collection of data is not so large their will be some error from the actual values obtained and that is why we can call every predicted value of any Random value as a 'Expected value' expressed by E( ) operator. The error can be minimizing by large sampling domain and by convenient. Sampling thus for a Random Variable X we can say $E(X)$ = expected mean; $E(X^2)$ = expected mean square. So on.

Chapter 3

INTRODUCTION TO SPECTRAL DENSITY

In a Layman's words we can say spectral density means Energy or Power density with respect to frequency in a signal i.e. spectral density=power or energy per unit of frequency.

Hence we can say TOTAL POWER or ENERGY in a SIGNAL=ESD or PSD×TOTAL frequency BW.ESD means Energy SPECTRAL DENSITY, PSD means POWER SPECTRAL DENSITY.

We know a periodic wave has infinite total energy and finite average power, hence for a periodic signal ESD is infinite and PSD is finite, similarly for an Aperiodic signal ESD is finite and PSD is zero. As we know a Random signal has Aperiodic orientation it also has infinite ESD (because of infinite coglomaration of aperiodic signal) but we can estimate theoretically an PSD of it.

We will express ESD of an signal x(t) as $\psi_x(j\omega)$ and PSD of a signal x(t) as $G_X(j\omega)$ respectively.

APERIODIC SIGNAL:

We know $E_T = \int_{-\infty}^{+\infty} \varphi_x(j\omega)df$, from the definition of $\varphi_x(j\omega)$ .Unit of $\varphi_x(j\omega)$ is joule/Hz

REAL SIGNAL	COMPLEX SIGNAL

$$E_T = \int_{-\infty}^{+\infty} x^2(t)\,dt$$

$$= \int_{-\infty}^{-\infty} x(t)x(t)\,dt$$

$$= \int_{-\infty}^{+\infty} x(t)[\int_{-\infty}^{+\infty} X(j\omega)e^{+j2\pi ft}\,df]\,dt$$

$$= \int_{-\infty}^{+\infty} X(j\omega)[\int_{-\infty}^{+\infty} x(t)e^{+j2\pi ft}\,dt]\,df$$

$$= \int_{-\infty}^{+\infty} X(j\omega)X(-j\omega)\,df$$

$$= \int_{-\infty}^{+\infty} X(j\omega)X*(j\omega)\,df$$

$$[X(-j\omega) = X*(j\omega)]$$

$$for a real signal$$

$$= \int_{-\infty}^{+\infty} |X(j\omega)|^2\,df$$

$$\therefore P_{av} = \int_{-\infty}^{+\infty} G_x(j\omega)\,df$$

$$= \underset{T \to \infty}{Lt}\, 1/T \int_{-T/2}^{+T/2} x^2(t)\,dt$$

$$= \underset{T \to \infty}{Lt}\, E_T/T = \underset{T \to \infty}{Lt}\, 1/T \int_{-\infty}^{+\infty} |X(j\omega)|^2\,df$$

$$= \int_{-\infty}^{+\infty} \underset{T \to \infty}{Lt}\, |X(j\omega)|^2 / T\,df$$

$$E_T = \int_{-\infty}^{+\infty} |x(t)|^2\,dt$$

$$= \int_{-\infty}^{+\infty} x(t)x*(t)\,dt$$

$$= \int_{-\infty}^{+\infty} x(t)[\int_{-\infty}^{+\infty} X*(j\omega)e^{-j2\pi ft}\,df]\,dt$$

$$= \int_{-\infty}^{+\infty} X*(j\omega)[\int_{-\infty}^{+\infty} x(t)e^{-j2\pi ft}\,dt]\,df$$

$$= \int_{-\infty}^{+\infty} X*(j\omega)X(j\omega)\,df$$

$$= \int_{-\infty}^{+\infty} |X(j\omega)|^2\,df$$

$$\therefore \varphi_X(j\omega) = |X(j\omega)|^2$$

$$\therefore G_X(j\omega) = \underset{T \to \infty}{Lt}\, |X(j\omega)|^2 / T$$
$$= 0$$

Analog Communication Engineering Fundamentals

PERIODIC SIGNAL:

$X(t) = 1/T_0 \sum\limits_{K=-\infty}^{+\infty} C_K e^{+jk\omega_0 t}$ (T_0 = time period)

REAL SIGNAL   COMPLEX SIGNAL

$P_{av} = 1/T_0 \int\limits_{-T_0/2}^{+T_0/2} x^2(t)dt$	$P_{av} = 1/T_0 \int\limits_{-T_0/2}^{+T_0/2} |x(t)|^2 dt$
$= 1/T_0^2 \sum\limits_{K=-\infty}^{K=\infty} C_K \int\limits_{-T_0/2}^{+T_0/2} x(t)e^{+jk\omega_0 t}dt$ [For a real signal $C_K^* = C_{-K}$]	$= 1/T_0 \int\limits_{-T_0/2}^{+T_0/2} x(t)x*(t)dt$
$= 1/T_0^2 \sum\limits_{K=-\infty}^{K=\infty} C_K C_{-K}$	$= 1/T_0^2 \sum\limits_{k=-\infty}^{\infty} C_K^* \int\limits_{-T_0/2}^{+T_0/2} x(t)e^{-jk\omega_0 t}dt$
$= 1/T_0^2 \sum\limits_{K=-\infty}^{\infty} C_K C_K^*$	$= 1/T_0^2 \sum C_K^* C_K$
$= \sum\limits_{K=-\infty}^{\infty} |C_K/T_0|^2$	$= \sum\limits_{K=-\infty}^{\infty} |C_K/T_0|^2$

Hence we have for a periodic signal

$P_{av} = \sum\limits_{K=-\infty}^{\infty} |C_K/T_0|^2$ This famous relation known as PARSEVAL's POWER THEOREM.

Now we know $\int\limits_{-\infty}^{+\infty} \delta(f - kf_0)df = 1$

So $P_{av} \times 1 = \sum\limits_{K=-\infty}^{\infty} |C_K/T_0|^2 \times \int\limits_{-\infty}^{+\infty} \delta(f - kf_0)df$

Or $\int\limits_{-\infty}^{+\infty} G_x(j\omega)df = \int\limits_{-\infty}^{+\infty} \sum\limits_{K=-\infty}^{+\infty} |C_K/T_0|^2 \delta(f - kf_0)df$

$$G_x(j\omega) = \sum\limits_{k=-\infty}^{+\infty} |C_K/T_0|^2 \delta(f - kf_0)$$

Now

$$P_{av} = \frac{Lt}{T \to \infty} E_T / T$$

$$or \frac{Lt}{T \to \infty} TP_{av} = E_T$$

$$or \int_{-\infty}^{+\infty} \frac{Lt}{T \to \infty} TG_x(j\omega)df = \int_{-\infty}^{+\infty} \varphi_x(j\omega)df$$

$$or \varphi_x(j\omega) = \frac{Lt}{T \to \infty} TG_x(j\omega)$$

$$= \frac{Lt}{T \to \infty} T \sum_{k=-\infty}^{+\infty} \left| C_K / T_0 \right|^2 \delta(f - kf_0)$$

hence $\psi_x(j\omega) = \infty$

PERIODIC SIGNAL	APERIODIC SIGNAL
$\psi_x(j\omega) = \infty$	$\left| X(j\omega)^2 \right|$
$G_x(j\omega) = \sum_{k=-\infty}^{+\infty} \left| C_K / T_0 \right|^2 \delta(f - kf_0)$	0

as $\delta(f - kf_0) = \delta(-f - kf_0)$, we can see

$$\psi_x(j\omega) = \left| X(j\omega) \right|^2 = \left| X(-j\omega) \right|^2 = \psi_x(-j\omega)$$

$$\& G_X(-j\omega) = \sum_{k=-\infty}^{+\infty} \left| C_{-K} / T_0 \right|^2 \delta(-f - kf_0) = G_X(j\omega)$$

Hence ESD & PSD's are even function of ω.

RELATION BETWEEN SPECTRAL DENSITIES & AUTOCORRELATION FUNCTION:

We know for aperiodic signal ESD $\psi_x(j\omega)$ for a aperiodic signal x(t) is $\psi_x(j\omega) = \left| X(j\omega)^2 \right|$ and

for a periodic signal x(t) the PSD $G_x(j\omega)$ is given as $G_x(j\omega) = \sum_{k=-\infty}^{+\infty} \left| C_K / T_0 \right|^2 \delta(f - kf_0)$; also

we know for aperiodic signal x(t),PSD $G_x(j\omega) = 0$ and for periodic signal ESD $\psi_x(j\omega) = \infty$.
As we know the fact periodic signal is power signal (with infinite total energy) and aperiodic signal is energy signal (with ZERO av. Power) .Let's find the relation between spectral densities and autocorrelation function ($\Re_x(\tau)$).

APERIODIC SIGNAL:

We know $\mathfrak{R}_x(\tau) = \mathfrak{R}_x(-\tau)$

Now $\mathfrak{R}_x(-\tau) = \int\limits_{-\infty}^{+\infty} x(t)x(t-\tau)dt = \mathfrak{R}_x(\tau)$

Let $x(-\tau) = V(\tau) = V[-(-\tau)]$ ie $x(\theta) = V(-\theta)$

Now $x(\tau) * x(-\tau) = x(\tau) * V(\tau) = \int_{-\infty}^{+\infty} x(t)V(\tau - t)dt$

Now $V(\tau - t) = V[-(t - \tau)] = x(t - \tau)$

Hence $x(\tau) * x(-\tau) = \int\limits_{-\infty}^{+\infty} x(t)x(t-\tau)dt = \mathfrak{R}_x(\tau)$.

Let assume $x(\tau) \xleftrightarrow{F} X(j\omega)$.

Hence $X(j\omega) = \int\limits_{-\infty}^{+\infty} x(\tau)e^{-j2\pi f \tau}d\tau$

Or $X(-j\omega) = \int\limits_{-\infty}^{+\infty} x(\tau)e^{+j2\pi f \tau}d\tau$

Substituting τ by- τ we have

$X(-j\omega) = \int\limits_{-\infty}^{+\infty} x(-\tau)e^{-j2\pi f \tau}d\tau$

Hence $X(-j\omega) \xleftrightarrow{F} x(-\tau)$

$F[x(\tau) * x(-\tau)] = X(j\omega)X(-j\omega)$

                 (from the correlation property of Fourier transform)

$F[\mathfrak{R}_x(\tau)] = |X(j\omega)|^2$ (as $\mathfrak{R}_x(\tau) = x(\tau) * x(-\tau)$)

$\therefore \mathfrak{R}_x(\tau) \xleftrightarrow{F} \psi_x(j\omega)$

Hence for an aperiodic signal Autocorrelation function & Energy spectral densities constitute Fourier transform pair.

PERIODIC SIGNAL:

Now for a periodic signal Time Autocorrelation function $\mathfrak{R}_x(\tau)$ is-

$\mathfrak{R}(\tau) = 1/T_0 \int\limits_{-T_0/2}^{+T_0/2} x(t)x(t+\tau)dt$

Now $x(t) = 1/T_0 \sum\limits_{k=-\infty}^{+\infty} C_k e^{+jk\omega_0 t}$

$x(t+\tau) = 1/T_0 \sum\limits_{m=-\infty}^{+\infty} C_m e^{+jm\omega_0(t+\tau)}$

now $\mathfrak{R}(\tau) = 1/T_0^3 \sum\limits_{m=-\infty}^{+\infty} C_m \sum\limits_{k=-\infty}^{+\infty} C_k e^{+jm\omega_0 \tau} \int_{-T_0/2}^{+T_0/2} e^{+j(k+m)\omega_0 t}dt$

$$=1/T_0^3 \sum_{K=-\infty}^{+\infty} C_K \sum_{m=-\infty}^{+\infty} C_m e^{+jm\omega_0\tau} \left[(2j\sin(k+m)\omega_0 T_0/2)/j(k+m)\omega_0 \right]$$

(now $\omega_0 T_0/2 = 2\pi/2 = \pi$)

$$=1/T_0^3 \sum_{K=-\infty}^{+\infty} C_K \sum_{m=-\infty}^{+\infty} C_m e^{+jm\omega_0\tau} \frac{2\sin(k+m)\pi}{(k+m)2\pi/T_0} \quad (\because \omega_0 = 2\pi/T_0)$$

$$=1/T_0^3 \times T_0 \sum_{K=-\infty}^{+\infty} C_K \sum_{m=-\infty}^{+\infty} C_m e^{+jm\omega_0\tau} \left[\sin(k+m)\pi/(k+m)\pi \right]$$

$$=1/T_0^2 \sum_{K=-\infty}^{+\infty} C_K \sum_{m=-\infty}^{+\infty} C_m e^{+jm\omega_0\tau} \text{ sinc } (k+m)$$

Now sinc (k+m) =0 for all (k+m); sinc (k+m) is always an integer.
And sinc (k+m) =1 for (k+m) =0 i.e. at k=-m.
Hence substituting m=-k and sinc (k+m) =1, we get

$$\Re(\tau) = 1/T_0^2 \sum_{K=-\infty}^{+\infty} C_K \sum_{k=-\infty}^{+\infty} C_K e^{-jk\omega_0\tau}$$

$$= \sum_{k=-\infty}^{+\infty} \left| C_K/T_0 \right|^2 e^{-jk\omega_0\tau}$$

Now $\Re(-\tau) = \sum_{k=-\infty}^{+\infty} \left| C_K/T_0 \right|^2 e^{+jk\omega_0\tau}$

Now we know $\Re_x(\tau) = \Re_x(-\tau)$;

So, $\Re(\tau) = \sum_{k=-\infty}^{+\infty} \left| C_K/T_0 \right|^2 e^{+jk\omega_0\tau}$ ....(1)

Now we know $F[e^{+jk\omega_0\tau}] = \int_{-\infty}^{+\infty} e^{-j(\omega-k\omega_0)\tau} d\tau$

$[\because \delta(f) = \int_{-\infty}^{+\infty} e^{-j\omega\tau} d\tau]$

$$= \int_{-\infty}^{+\infty} e^{-j2\pi(f-kf_0)\tau} d\tau = \delta(f-kf_0)$$

or $\Re(\tau) == \sum_{k=-\infty}^{+\infty} \left| C_K/T_0 \right|^2 F^{-1}[\delta(f-kf_0)]$

$$= F^{-1}[\sum_{k=-\infty}^{+\infty} \left| C_K/T_0 \right|^2 [\delta(f-kf_0)]]$$

$$\because [\sum_{k=-\infty}^{+\infty} \left| C_K/T_0 \right|^2 [\delta(f-kf_0)]] = G_X(j\omega) \text{ (for periodic signal)}$$

Or, $\Re(\tau) = F^{-1}[G_X(j\omega)]$

Or $F[\Re(\tau)] = G_X(j\omega)$

Analog Communication Engineering Fundamentals

$\therefore \Re_{xx}(\tau) \xleftarrow{\ F\ } G_X(j\omega)$.

Hence autocorrelation function constitutes a Fourier transform pair with power spectral density for a periodic signal.

POWER SPECTRAL DENSITY OF A RANDOM SIGNAL:
Already we have determined spectral density for periodic and aperiodic signal.
Now let us try to calculate spectral density for a random signal.
Consider a sample function of a random signal.

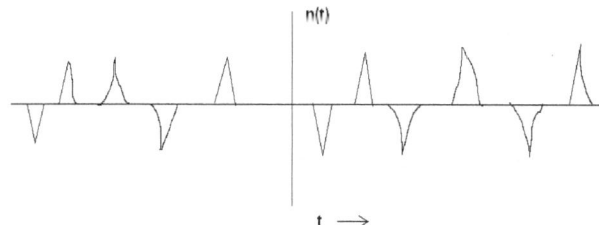

We consider the following conditions:

1) The pulses are the same form, but have random amplitude and statistically independent random types of occurrences.
2) The waveform is stationary, so that the stationary features of wave forms(i.e. the process) time invariant.
3) There is an invariant average time of separation T_s between pulses.
4) There are no overlaps between pulses.

Consider A Pulse:
If the Fourier Transform of a signal sample pulse $P_1(t)$ is $P_1(j\omega)$.then from PARSEVAL's theorem we have the normalized energy of the pulse is $E_1 = \int_{-\infty}^{+\infty} P_1(j\omega) \cdot P_1(-j\omega)df$

$$= \int_{-\infty}^{+\infty} |P_1(j\omega)|^2 \, df .$$

So the energy in the range df at a frequency f is $dE_1 = |P_1(j\omega)|^2 \, df$.

Let consider n successive pulses. Since we have assumed that pulses do not overlap the energy in the range df at the frequency f due to n pulses is $dE = dE_1 + dE_2 + \ldots\ldots\ldots + dE_n$

$$= [|P_1(j\omega)|^2 + |P_2(j\omega)|^2 + \ldots\ldots\ldots + |P_n(j\omega)|^2] \, df$$

the mean value $\overline{|P(j\omega)|}^2$ of the sequences of pulses of n pulses is-

$$\overline{|P(j\omega)|^2} = 1/n[\ |P_1(j\omega)|^2 + |P_2(j\omega)|^2 + \ldots\ldots + |P_n(j\omega)|^2]$$

so, $dE = n \ \overline{|P(j\omega)|^2} \ df$

we have assumed that the average time of separation between pulses is T_s. so that n pulses will occur in a time gap of nT_s.

so the power content in the band df is

$P_{df} = dE/T_s = 1/T_s \ \overline{|P(j\omega)|^2} \ df.$

Let $T_s = T$ and $P(j\omega) = X(j\omega)$ and $T \to \infty$

Now $G_x(j\omega) = p_{df}/df = $ power spectral density

i.e. $G_x(j\omega) = \underset{T \to \infty}{Lt} \ 1/df.1/T \ \overline{|X(j\omega)|^2} \ df$

$\therefore \ G_x(j\omega) = \underset{T \to \infty}{Lt} \ E\overline{|X(j\omega)|^2} /T.$

Now spectral density $G_x(j\omega)$ for a random signal denotes as $S_x(j\omega)$ to distinguish it from it's deterministic counter parts as in deterministic case

$$R_{XX}(\tau) \quad \overset{f}{\longleftrightarrow} \quad S_x(j\omega)$$

(this proof is out far out present scope)

$$S_x(j\omega) = \int_{-\infty}^{+\infty} R_{XX}(\tau)e^{-j2\pi f\tau} d\tau$$

and $R_{XX}(\tau) = \int_{-\infty}^{+\infty} S_x(\tau)e^{+j2\pi f\tau} df$

or $S_x(0) = \int_{-\infty}^{+\infty} R_{XX}(\tau) \ d\tau$ =total area under autocorrelation curve.

$R_{XX}(0) = \int_{-\infty}^{+\infty} S_x(j\omega)df$ =total power of random signal

$\Rightarrow R_{XY}(\tau) = E[X(t)Y(t+\tau)]$

$\quad R_{XX}(\tau) = E[X(t)X(t+\tau)]$

$\quad\quad R_{XX}(0) = E[X^2(t)] = \int_{-\infty}^{+\infty} S_x(j\omega)df$.

All properties of $R_{XY}(\tau)$ and $R_{XX}(\tau)$ are similar to $\Re_{xy}(\tau)$ and $\Re_{xx}(\tau)$

Just replace( ) operator by the non deterministic mean or Expectation operator E( ).

Now ,

$\quad\quad S_x(j\omega) = \underset{T \to \infty}{Lt} \ E\overline{|X(j\omega)|^2} /T.$

$$S_X(-j\omega) = \underset{T \to \infty}{Lt} \quad E\overline{|X(-j\omega)|}^2 / T$$

$$= \underset{T \to \infty}{Lt} \quad E\overline{|X(j\omega)|}^2 / T = S_X(j\omega).$$

Hence as it's deterministic counter part $S_X(j\omega)$ is also even function of ω.

Hence $\int\limits_{-\infty}^{+\infty} G_X(j\omega) or \ S_X(j\omega) or \psi_x(j\omega) \, df$

$$= 2\int\limits_{0}^{+\infty} G_X(j\omega) or \ S_X(j\omega) \ or \psi_x(j\omega) \, df$$

[Sometimes the term COHERENCE used in place of CORRELATION].

Chapter 4

SAMPLING ANALYSIS

What is SAMPLING?

Sample means a specimen. Sample of a electrical signal means a specimen of that electrical signal obtained in a fixed time gap or at a well defined time spot.

PURPOSE OF SAMPLING:- By sampling we cam convert a continuous signal into a discrete signal, a discrete signal into a more discrete signal. Sampling Analysis reveals the fact (i.e. Sampling Theorem) that by means of proper sampling we can RECOVER the original MESSAGE from it's sampled version. So implementing the above criteria (i.e. Sampling Technique) we can design a system which consume LESS POWER (due to discrete transmission or inspection of an continuous signal), which MULTIPLEX (i.e. many –into-one, because one signal need not be observe continuously) several signals in one channel and so increasing the efficiency of communication.

SAMPLING ANALYSIS:- In Sampling Analysis we will study the nature & space of sampling of both continuous & discrete signals, processing of that sampled version of signals and finally the recovery principle of original message from sampled version. Sampling Analysis consists of two section-1) SAMPLING THEOREM; here we will study mathematical background and related spectrum analysis with Nyquist Theorem and 2) SAMPLING TECHNIQUE; here we will study DESIGN and Behavior of circuitry which are employed to implement Sampling process i.e. sampling of message signal, processing of sampled signal, Recovery of message signal etc.

SAMPLING THEOREM

Sampling Theorem's consists of FOUR STATEMENTS, out of these the FIRST and the LAST are very very important:-

STATEMENT 1 :- (Nyquist Theorem/ SAMPLING PRINCIPLE) A SIGNAL BANDLIMITED to B Hz (i.e. the signal has no frequency components at $|f|>B$) is UNIQUELY DETERMINED BY IT'S VALUES SAMPLED AT UNIFORM INTERVALS LESS THAN or EQUAL 1/2b Seconds APART.

Proof:-
A) Continuous Signal:-

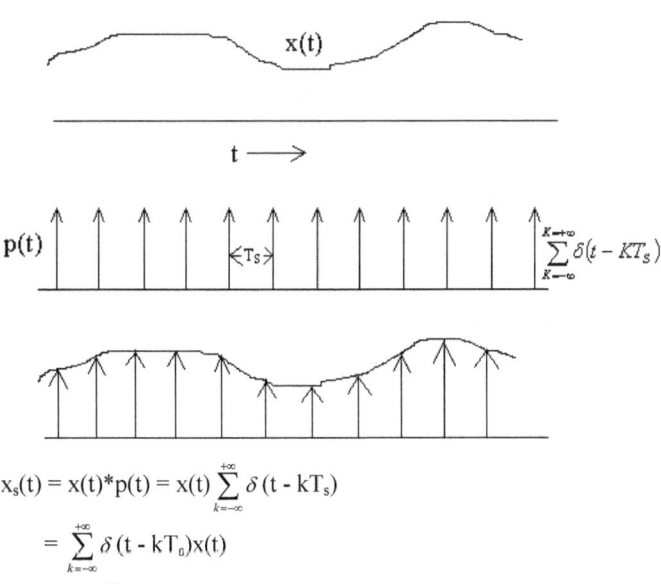

$$x_s(t) = x(t)*p(t) = x(t) \sum_{k=-\infty}^{+\infty} \delta (t - kT_s)$$

$$= \sum_{k=-\infty}^{+\infty} \delta (t - kT_0)x(t)$$

$$\therefore x_s(t) = \sum_{k=-\infty}^{+\infty} \delta (t - kT_s)x(kT_s)$$

Let $x(t)$ is a continuous signal which frequency spectrum (i.e. Fourier Transform) is $x(j\omega)$ as shown in figure. This continuous signal $x(t)$ is sampled by a SAMPLING PULSE (Ideal Sampling , Sampling PULSE IS impulse

TRAIN) of p(t) ,where p(t) is $p(t) = \sum\limits_{k=-\infty}^{+\infty} \delta(t - kT_s)$; i.e. p(t) is a train of IMPULSE . Now if we MULTIPLY x(t) by p(t) the resultant is our required sampled version of x(t) i.e. $x_s(t)$

where $x_s = \sum\limits_{k=-\infty}^{+\infty} \delta(t - kT_s) \, x(kT_s)$

where k is an integer , T_s is the PERIOD of Sampling Impulse i.e. $f_s = 1/T_s$ is the sampling frequency , impulse train p(t) sampled the value of x(t) at time $t = T_s$ and at every integer multiple of T_s; now we wish to determined the frequency spectrum of $X_s(t)$-

so $F[X_s(t)] = X_s(j\omega) = [\sum\limits_{k=-\infty}^{+\infty} \delta(t - kT_s) \, x(kT_s)]$-------------------(1)

Now we know for a periodic PULSE train of pulse width τ and with time period T_0, $g(t) = A_\tau/T_0 \sum\limits_{k=-\infty}^{+\infty} [\sin k\omega_0(\tau/2) / k\omega_0(\tau/2)] e^{+jk\omega_0 t}$

where A = pulse amplitude

now if $\tau \to 0$ keeping $A_\tau = 1$ and due to fact $\mathop{Lt}\limits_{\theta \to 0}(\sin\theta/\theta)=1$, then if $T_0 = T_s$ then

periodic pulse train become impulse train i.e. $\mathop{Lt}\limits_{\tau \to 0} g(t) = p(t)$

or, $\mathop{Lt}\limits_{\tau \to 0} A_\tau/T_0 \sum\limits_{k=-\infty}^{+\infty} [\sin k\omega_0(\tau/2) / k\omega_0(\tau/2)] e^{+jk\omega_0 t} = p(t)$

or, $p(t) = 1/T_s \sum\limits_{k=-\infty}^{+\infty} e^{+jk\omega_0 t}$ ($\because T_0 = T_s$ then $\omega_0 = \omega_s$)

or, $\sum\limits_{k=-\infty}^{+\infty} \delta(t - kT_s) = 1/T_s \sum\limits_{k=-\infty}^{+\infty} e^{+jk\omega_0 t}$

Substituting the above in equation (1), we get –

$X_s(j\omega) = F[1/T_s \sum\limits_{k=-\infty}^{+\infty} x(kT_s) e^{+jk\omega_s t}]$

$= 1/T_s \sum\limits_{k=-\infty}^{+\infty} F[x(kT_s) e^{+jk\omega_s t}]$

$= 1/T_s \sum\limits_{k=-\infty}^{+\infty} \int_{-\infty}^{+\infty} x(kT_s) e^{-j(\omega - k\omega_s)t} \, dt$

$= 1/T_s \sum\limits_{k=-\infty}^{+\infty} \int_{-\infty}^{+\infty} x(t) e^{-j(\omega - k\omega_s)t} \, dt$

$= 1/T_s \sum\limits_{k=-\infty}^{+\infty} \int_{-\infty}^{+\infty} x[j(\omega - k\omega_s)]$

$[\because x(t)\delta(t-kT_s)= x(kT_s)\ \delta(t-kT_s)]$; see the figure ; we can write x(t) in place of x(kT_s) or we can simply start from $x_s(t) = \sum\limits_{k=-\infty}^{+\infty} x(t)\ \delta(t-kT_s)]$

Hence we get $X_s(j\omega) = 1/T_s \sum\limits_{k=-\infty}^{+\infty} x[j(\omega-k\omega_s)]$

i.e. in spectrum of (observe that in the expression of $X_s(j\omega)$ only ω is variable , k & ω_s are constant) $X_s(j\omega)$) $X(j\omega)$ is repeatedly occur for
k = 0, ± 1, ± 2, ± 3---------- at the spacing of ω_s which is sampling frequency itself , so from this conclusion we can plot the spectrum of $X_s(j\omega)\rightarrow$
(only k = 0 & k =1 spectrum are shown)

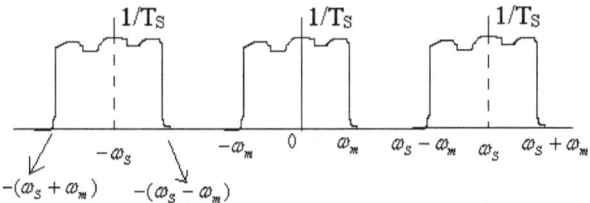

Now it is clear that if we want to recover $X(j\omega)$ from $X_s(\omega)$ adjacent replicas of $X(j\omega)$ should not be overlapped in $X_s(j\omega)$ i.e. $\omega_s - \omega_m$; should greater or at least equal to ω_m; so for faithfully recovery of message we need

$$\omega_s - \omega_m \geq \omega_m$$
or, $\omega_s \geq 2\omega_m$
$\therefore f_s \geq 2f_m$ \qquad (here f_m = maximum frequency component present in message = B)

& $T_s \leq \frac{1}{2} f_m$
$\therefore T_s \leq \frac{1}{2}B$ (proved)

B) <u>DISCRETE SIGNAL</u> :- Consider a DISCRETE SIGNAL X[n] and it's fourier transform (i.e. frequency spectra) $X(j\omega)$. Again consider the same IMPULSE train p[n] (not p(t) here) for sampling purpose. Multiplying X[n] by P[n] we can produce sampled version of X[n] as $X_s[n]$.

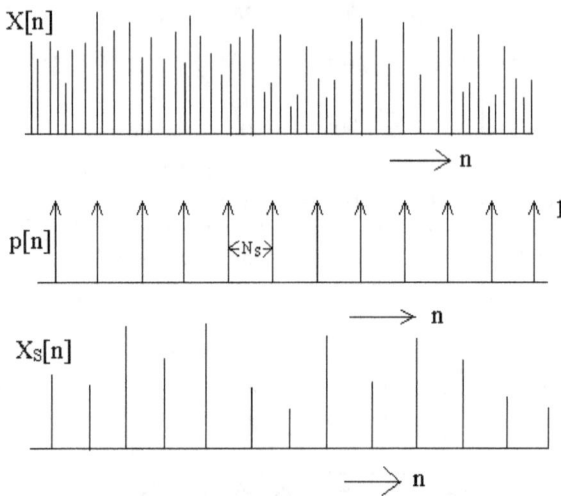

Let N_s is the discrete sampling interval (similar to T_s in continuous sampling interval , here although $N_s = 1$, we have to write N_s instead of T_s as sampling interval to represent the discrete case. But to calculate ω_s as a discrete frequency we have to calculate T_s or actual time gap between sampling instant because as $N_s = 1$ bears no separation information here) hence $\omega_s = 2\pi/N_s$ and p[n] will be p[n]

$$= \sum_{k=-\infty}^{+\infty} \delta[n - kN_s]$$

Hence the sampled version of X[n] , obviously $X_s = X[n]\, p[n]$

Hence the spectrum of $X_s[n]$ is $X_s[j\omega]$ and

$$X_s[j\omega] = F(X[n]\, p[n])$$

$$= F(X[n] \sum_{k=-\infty}^{+\infty} \delta[n - kN_s])$$

$$\therefore\ X_s[j\omega] = F(\sum_{k=-\infty}^{+\infty} X[kN_s]\, [\delta(n-kN_s)]\text{-----------------------------}(2)$$

MATHEMATICAL REVIEW :-

(A) $\sum_{n=0}^{r} \alpha^n = (r+1)$, if $\alpha = 1$

Proof :- if $\alpha = 1$, then

$$\sum_{n=0}^{r} \alpha^n = \sum_{n=0}^{r} (1)^n = (1 + 1 + 1 + \text{---------}(r+1)\text{times})$$

$$= r + 1 \text{ (proved)}$$

(B) $\sum_{n=0}^{r} \alpha^n = \dfrac{1-\alpha^{(r+1)}}{1-\alpha}$, if α is any complex number $\neq 1$

Proof :- Let $S = \sum_{n=0}^{r} \alpha^n$

so, $\alpha S = \sum_{n=0}^{r} \alpha^{n+1}$

so, $S - \alpha S = \sum_{n=0}^{r} \alpha^n - \sum_{n=0}^{r} \alpha^{n+1}$

or, $S(1 - \alpha) = \sum_{n=0}^{r} \alpha^n - (\sum_{n=0}^{r} \alpha^n - 1 + \alpha^{(r+1)})$

or, $S(1 - \alpha) = 1 - \alpha^{(r+1)}$

or, $S = \dfrac{1-\alpha^{(r+1)}}{1-\alpha}$

$\therefore \sum_{n=0}^{r} \alpha^n = \dfrac{1-\alpha^{(r+1)}}{1-\alpha}$ [For α is any complex number $\neq 1$](proved)

Now consider a discrete pulse train of period N_0 and pulse width N_r with amplitude A as follows.

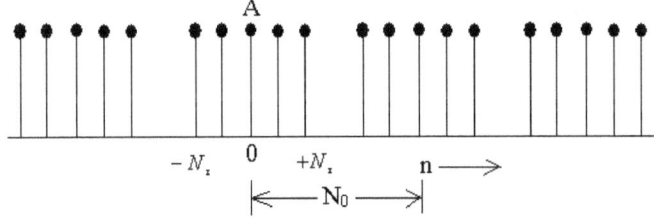

Pulse Train g[n]

Now from our previous knowledge of discrete time Fourier series, Fourier coefficient of above g[n] will be

$$a_k = A \frac{A}{N_0} \frac{Sin[2\pi k(N_r + 0.5)/N_0]}{Sin(\pi k/N_0)}$$

when $k \neq 0, \pm N_0, \pm 2N_0 \text{-------------}$

$$= \frac{A(2N_r + 1)}{N_0}$$

Hence we know $g[n] = \sum\limits_{k=<N_0>} a_k\, e^{jk(2\pi/N_0)n}$ when $k \neq 0, \pm N_0, \pm 2N_0 \text{------------}$

Now look more close at g[n].If we letting $N_r \longrightarrow 0$,g[n] becomes more and more like an impulse train and at limit it becomes a impulse train; hence if we consider $N_0 = N_S$,

A = 1 then $\underset{N_r \to 0}{Lt}\, g[n] = p[n]$

Or, $p[n] = \underset{N_r \to 0}{Lt} \sum\limits_{k=N_s} a_k e^{jk(2\pi/N_s)n}$

Considering A = 1 we get $\rightarrow (\omega_s = 2\pi/N_s)$

$P[n] = 1/N_s \sum\limits_{k=N_s} (\underset{N_r \to 0}{Lt}\, a_k) e^{jk\omega_s n}$

Observe from the equation of a_k in both case

$\underset{N_r \to 0}{Lt}\, a_k = 1/N_s \quad [\because N_0 = N_s, A = 1]$

or, $p[n] = 1/N_s \sum\limits_{k=N_s} e^{jk\omega_s n}$

now from equation (2) we know-

$X_s[j\omega] = F(x[n]p[n])$

$= F(x[n]1/N_s \sum\limits_{k=N_s} e^{jk\omega_s n})$

$= 1/N_s \sum\limits_{k=N_s} F(x[n]e^{jk\omega_s n})$

$= 1/N_s \sum\limits_{k=N_s} (\sum\limits_{n=-\infty}^{+\infty} x[n]e^{-j(\omega - k\omega s)n})$

$= 1/N_s \sum\limits_{k=N_s} X[j(\omega - k\omega s)]$

$\therefore X_s[j\omega] = 1/N_s \sum\limits_{k=N_s} X[j(\omega - k\omega s)]$

Here a difference occurs in this expression compared to dual expression of X_s $(j\omega)$. For continuous case the difference occurred in the limit of K.But this is not an amazing case, we know that discrete periodic signal is periodic also in frequency domain by a frequency interval of 2π rad/s.There is only N_0 different no. of harmonics present ($N_0 = \dfrac{2\pi}{\omega_0}$).So one have to include only N_s no. of harmonics in discrete sampling, but this is not the case for continuous signal. So then the range of K will be $-\infty \rightarrow +\infty$ etc.

Now similarly we can draw the spectrum of $X_s[j\omega]$ as follows:-

Look the difference between the spectra of $X_s[j\omega]$ and $X_s[j\omega]$. Spectrum of $X_s[j\omega]$

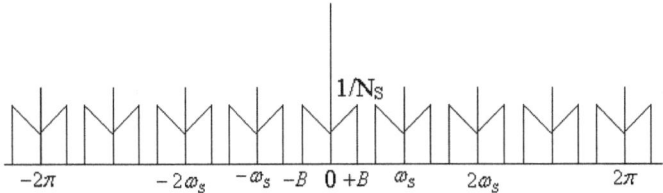

is limited within the frequency range of 2π rad/s. But the spectra of $X_s[j\omega]$ is continuous by unlimitedly.

Now similarly for faithful recovery of the signal we need-

$$\omega_s - B(2\pi) \geq B(2\pi)$$
$$or, f_s - B \geq B$$
$$or, f_s \geq B$$
$$\therefore N_s \leq 1/2B$$

(proved)

so, at least we can conclude by saying that any signal which has no frequency components greater than B, can be reproduced accurately if we sample that signal with the sampling frequency $f_s \geq 2B$ or by spacing the sampling instant at uniform time gap of $T_s(N_s) \leq 1/2B$.

The above sampling principle is called NYQUIST THEOREM. 2B is called Nyquist Frequency (also Nyquist rate) and 1/2B is called Nyquist time gap.

STATEMENT 2(Sampling Rate):-

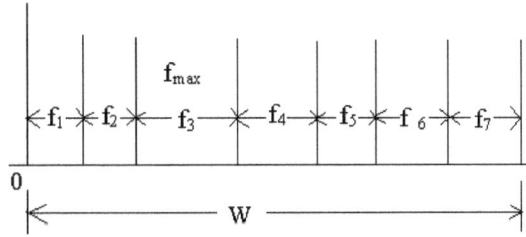

IF A TIME FUNCTION x(t) consists of a BAND OF FREQUENCIES displaced from ZERO, with a total bandwidth of w and with highest frequency f_{max}, then the minimum sampling rate ia given by $2f_{max}/m$ where m is the largest integer not to exceed f_{max}/w.

Proof:-

Consider the above diagram of frequency spectra. We can consider that it's a spectra of a signal which has no frequency component greater than w. Hence the Nyquist frequency is 2w. Now according to theorem $2w = 2f_{max}/m$. $\therefore m = f_{max}/w$ i.e. if we replaced m by f_{max}/w in the ratio of $2f_{max}/m$ we get back 2w which is Nyquist rate. This agree with STATEMENT 1 and hence completes the proof. Observe for a single frequency $f_{max}=w$ and $m=1$ Nyquist rate will be $2f_{max}$ as we find carrier.

STATEMENT 3(Frequency Domain Specification):-

In the statement 1 we have assumed that signal is Band limited. Now for a time limited signal g(t) which is ZERO except in the range of $T_1 \le t \le T_2$ G(f) or frequency spectra of g(t) can be completely determined by specifying it's values at s series of points spaced every $1/T_2$-T_1 Hz.
Proof:-

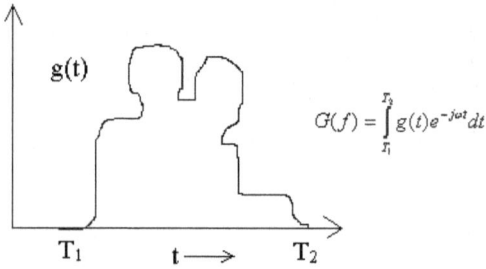

$$G(f) = \int_{T_1}^{T_2} g(t)e^{-j\omega t}dt$$

$T_1 \qquad t \longrightarrow \qquad T_2$

If we assume that above APERIODIC waveforms g(t) as PERIODIC wave of time period $(T_2 - T_1)$ then one can express g(t) in FOURIER SERIES expression

$$g(t) = 1/T_2\text{-}T_1 \sum_{k=-\infty}^{+\infty} C_k e^{+j2\pi kt} /(T_2 - T_1)$$

$$\text{where } C_k = \int_{T_1}^{T_2} g(t)e^{-j2\pi kt} /(T_2 - T_1)\, dt$$

$$= \int_{T_1}^{T_2} g(t)e^{-j2\pi(k/T_2 - T_1)t})$$

$$= G(k/T_2\text{-}T_1)$$

now if we able to determined C_k we can determine g (t) completely. Now we get C_k = $G(k/T_2 - T_1)$ i.e. the frequency spectra of that signal need only to be sample at every sampling point in frequency domain at an interval of $(k/T_2 - T_1)$

STATEMENT 4(Signal Reconstruction):-

If a signal with highest frequency B has been sampled a rate of 2B sample/S (i.e. at Nyquist Rate), and the samples are in the form of impulse whose area is proportional to the magnitude of the sample at that instant then the sampled signal maybe reconstructed by passing the impulse which through an IDEAL LOW-PASS FILTER with cut off frequency of B.
Of B.

Proof :-
To proof the above statement, one can simply look at the spectra of $X_s(j\omega)$ and clearly understand that the $X_s(j\omega)$ is required to pass through a LPF at gain T_s (or N_s) with a pass band of $B \leq$ LPF Pass band$<f_s$-B or the Transfer function of such a filter is $H(j\omega) = T_s\ \pi(f/2B)$.
But this is not a mathematical proof although complicated one has to prove mathematically that we need a LPF for recovery of the signal. Here goes let for a signal (which is our message here).
We already familiar with the following equations$\rightarrow$

$$X_s(t) = \sum_{k=-\infty}^{+\infty} x(kT_s)\delta(t - kT_s)\text{-----------------------------(A)}$$

$$X_s(j\omega) = 1/T_s \sum_{k=-\infty}^{+\infty} X[j(\omega - k\omega_s)]\text{-----------------------(B)}$$

Now in the range of $-2\pi B \leq \omega \leq 2\pi B$, $X_s(j\omega) = X(j\omega)/T_s$, if we choose $T_s = 1/2B$ exactly (i.e. Nyquist interval) then considering the above fact from equation (B) we can write-

$$X(j\omega) = 1/2B\ X_s(j\omega), \qquad -B \leq f \leq B\text{-----------------------(C)}$$

Now if fourier transform both the sides of equation (A) we get-

$$X_s(j\omega) = F[\sum_{k=-\infty}^{+\infty} x(kT_s)\delta(t - kT_s)]$$

$$= \sum_{k=-\infty}^{+\infty} x(kT_s)F[\delta(t - kT_s)]$$

as we know $\delta(t - t_0) \Leftrightarrow e^{-j2\pi ft_0}$

$$\therefore X_s(j\omega) = \sum_{k=-\infty}^{+\infty} x(kT_s)e^{-2\pi fkT_s}\text{------------------------------------(D)}$$

Hence combining equation (C) & (D) we can write-

$$2BX(j\omega) = \sum_{k=-\infty}^{+\infty} x(kT_s)e^{-2\pi fkT_s} \qquad -B \leq f \leq B$$

as we considered already $T_s = 1/2B$-

hence, $\quad X(j\omega) = 1/2B \sum\limits_{k=-\infty}^{+\infty} x(k/2B)e^{-j2\pi kf/B}$, $\quad -B \le f \le B$

now $x(t) \Leftrightarrow X(j\omega)$

so $x(t) = \int\limits_{-\infty}^{+\infty} X(j\omega)e^{+j2\pi ft}\,df = \int\limits_{-B}^{+B} X(j\omega)e^{+j2\pi ft}\,df$

or, $x(t) = \int\limits_{-B}^{+B} 1/2B \sum\limits_{k=-\infty}^{+\infty} x(k/2B)\,e^{[j2\pi f(t-k/2B)]}\,df$

$\qquad = 1/2B \sum\limits_{k=-\infty}^{+\infty} x(k/2B) \int\limits_{-B}^{+B} \exp[j2\pi f(t-k/2B)]\,df$

Let $\theta = 2\pi f(t-k/2B)$

Or $d\theta/df = 2\pi(t-k/2B)$

When $f = \pm B$, $\theta = 2\pi(\pm B)(t-k/2B) = \pm 2\pi Bt \mp \pi k$

So, $x(t) = 1/2B \sum\limits_{k=-\infty}^{+\infty} x(k/2B) \int\limits_{-2\pi Bt+k\pi}^{2\pi Bt-k\pi} e^{j\theta}\,d\theta(\dfrac{1}{2\pi t - \pi k/B})$

$\qquad = \sum\limits_{k=-\infty}^{+\infty} x(k/2B)\dfrac{Sin(2\pi Bt - k\pi)}{(2\pi Bt - k\pi)}$

$\qquad = \sum\limits_{k=-\infty}^{+\infty} x(k/2B)\,Sinc(2Bt-k), \quad -\infty \le t \le \infty$ --------------(E)

Now from the above conation we have whenever $t = rT_s$ (where k=r an integer &

we know $T_s = 1/2B$)then $x(t) = x(r/2B)$ [$\because \sin(0) = 1$] i.e. at a very $t =$

$kT_s(-\infty \le k \le +\infty)$ point the above equation gives $x(t) =$ sampled value of $x(t)$ at that

point. Out come of the above equations are conglomeration of 'sinc' curve each

has it's peak value at the corresponding values of K; between the two values of

K(i.e. between two integer) the curve of $x(t)$ approximated by the value of

precursor and post cursor by the several approximation method. These

phenomenon known as CURVE FITTING or INTERPOLATION because the

approximination method Interpolate between the two values of $X(t)$. The simplest interpolation or approximation method is linear interpolation Higher order interpolations are also possible. Equation (E) is called interpolation equation. From our previous knowledge we know the equation $x(t_k) = x(k/2B)\text{Sinc}(2Bt - k)$ is the time domain response at the following frequency spectra weighted by $x(k/2B)$.clearly that is the frequency spectra of a LOW-PASS FILTER now if the sum the above response for $k = -\infty$ to $+\infty$ then clearly according to interpolation equation (i.e. equation E) a LPF will reconstructed the message that is why LOW-PASS filter often call a interpolating filter. This completes the proof of statement 4.

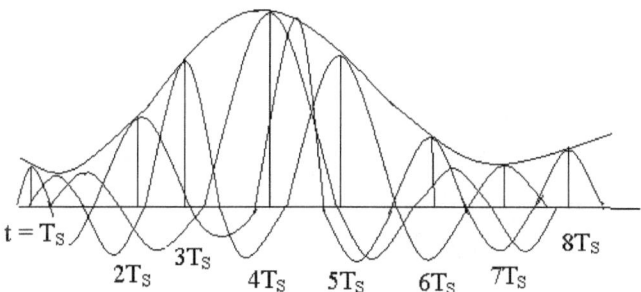

Chapter 5

Amplitude Modulation

What is AM?
In an AM system the amplitude of a radio frequency signal called the carrier is made proportional to the instantaneous amplitude of the message signal called the modulating signal.

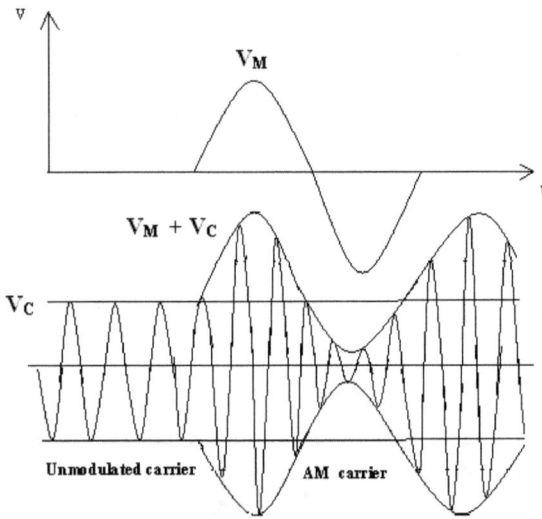

Let modulating waves (signal) V_m (t) = $E_m \cos \omega_m t$ and carrier waves V_c (t) = $E_c \cos \omega_c t$ and $\omega_c \gg \omega_m$

We have ignored the phase for simplicity.

Frequency Spectra of AM Waves

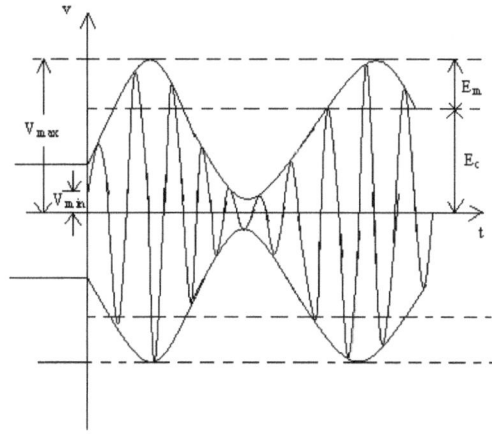

page 2

Now the envelop of AM wave is –

E = max^m amplitude of carrier $+V_m$

$= E_c + E_m \sin \omega_m t$

$= E_c (1 + E_m/E_c \sin \omega_m t)$

$= E_c (1 + k_a \sin \omega_m t)$ [Let $k_a = E_m/E_c$]

So the AM wave is-

$V(t) = E \sin \omega_c t = E_c(1 + k_a \sin \omega_m t) \sin \omega_c t$

Or, $V(t) = E_c(1 + k_a \cos \omega_m t) \cos \omega_c t$--------------------------------------(1)

$m \leq 1$, known as modulation index.

Percentage of $k_a = E_m/E_c \times 100\%$. (Also depth of modulation)

Expression (1) is AM wave for single modulating signal called TONE MODULATION. Now if we take modulating signal frequency as $V_m(t)$ i.e. $V_m(t)$ or simply $m(t) = \cos \omega_m t$

Then $V(t) = E_c(1 + k_a m(t)) \cos \omega_c t$

   $\therefore V(t) = E_c(1 + k_a m(t)) \cos \omega_c t$ [k_a = modulating index]

This $E_c(1 + k_a m(t))$ = amplitude of AM wave

Now as $E_m/E_c \langle 1$ so, $k \langle 1$ so, $-1 \langle k_a m(t) \langle 1$

Hence $|k_a m(t)| \langle 1$ as amplitude of $m(t)$ here varies between +1 to -1

So $1 + km(t) \rangle 0$ and $1 + k_a m(t) \langle 2$

i.e. $0 \langle 1+k_am(t) \langle 2$

Now if $|k_a m(t)| \geq 1$ then for some m(t) when $k_am(t) \langle -1$, $1+k_am(t) \langle 0$ or the amplitude of AM wave i.e. of V(t) is negative and ENVELOPE DISTORTION will occur.

So the two requirements of proper AM operation is-

1. Carrier frequency f_c is much greater than the highest frequency component W of the message signal m(t) , i.e.

$$f_c \rangle \rangle W$$

Obviously W is the message bandwidth . If this condition is not satisfied an envelope cannot be visualized satisfactory.

2. $|k_a m(t)| \langle 1$ i.e. $1+k_am$ (t) should always be positive. Otherwise Envelope distortion can occur.

$$g(t) \leftrightarrow D(f) \qquad D(t) \leftrightarrow g(-f)$$

now let $m(t) \leftrightarrow M(f)$

or $M(f) = \int_{-\infty}^{+\infty} m(t)e^{-j2\pi ft}dt$

Now $\cos(w_ct) = \frac{1}{2}(e^{-jw_ct} + e^{+jw_ct})$

So $m(t) \cos(w_ct) = \frac{1}{2}e^{-jw_ct}m(t) + e^{+jw_ct}m(t)$

So $F[m(t)\cos w_ct] = \frac{1}{2}\int_{-\infty}^{+\infty} m(t)e^{-j2\pi(f+f_c)t}dt + \frac{1}{2}\int_{-\infty}^{+\infty} m(t)e^{-j2\pi(f-f_c)t}dt$

$$= \frac{1}{2}M(f+f_c) + \frac{1}{2}M(f-f_c)$$

and $F[\cos w_ct] = \frac{1}{2}\int_{-\infty}^{+\infty} e^{-j2\pi(f+f_c)t}dt + \frac{1}{2}\int_{-\infty}^{+\infty} e^{-j2\pi(f-f_c)t}dt$

$$= \frac{1}{2}\delta(f+f_c) + \frac{1}{2}\delta(f-f_c) \qquad [\because \delta(f) = \int_{-\infty}^{+\infty} e^{-j2\pi ft}dt]$$

Now $V(t) = E_c(1+k_am(t))\cos \omega_c t$

$F[V(t)] = E_c F[\cos \omega_c t] + k_aE_c F[m(t)\cos \omega_c t]$

$$= E_c/2 \, \delta(f+f_c) + k_aE_c/2 \, M(f+f_c) + E_c/2 \, \delta(f-f_c) + k_aE_c/2 \, (f-f_c)$$

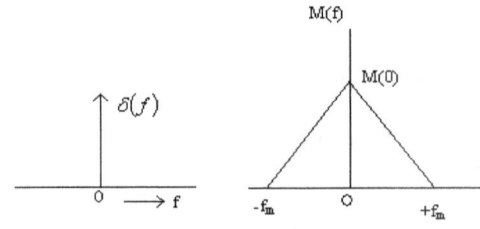

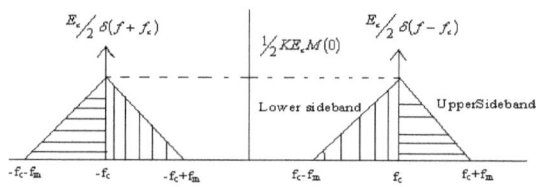

Another look

We get, $V(t) = E_c (1+k_a\cos\omega_m t)\cos\omega_c t$

$\qquad = E_c\cos\omega_c t + k_aE_c/2 . 2\cos\omega_m t \cos\omega_c t$

$\qquad = E_c \cos\omega_c t + k_aE_c/2[\cos(\omega_m+\omega_c)t + \cos(\omega_c - \omega_m)t]$

$\qquad = E_c \cos\omega_c t + k_aE_c/2\cos(\omega_m+\omega_c)t + k_aE_c/2 \cos(\omega_c - \omega_m)t$

Let $V(t) \leftrightarrow V(f)$

So $V(f) = E_c/2\, \delta(f\pm f_c) + k_aE_c/2\, \delta[f\pm(f_c+f_m)] + k_aE_c/2\, \delta[f\pm(f_c-f_m)]$

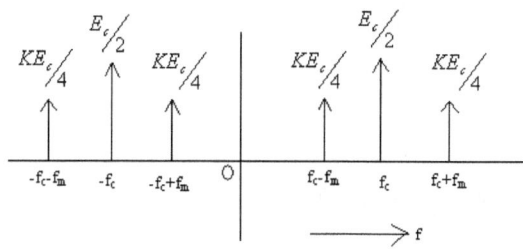

So in both analysis we get

$\qquad V(t) = $ Carrier $+$ USB $+$ LSB

TOTAL POWER in AM waves:-

$\qquad$ Power in $V(t) = P_t$

$\qquad P_t = P_c + P_{USB} + P_{LSB}$ $\qquad\qquad$ Now $P_{USB} = P_{LSB} = P_{SB}$

$\qquad\quad = P_c + 2P_{SB}$

$\qquad\quad = (E_c/\sqrt{2})^2 + 2.((k_aE_c/2)/\sqrt{2})^2$

$\qquad\quad = E_c^2/2 + k_a E_c^2/4$

$\qquad \therefore P_t = (1+k_a^2/2) E_c^2/2 = (1+k_a^2/2)P_c$

MULTI FREQUENCY AM

Let there be multiple modulating frequencies f_1, f_2, f_3--------- and for each of them the modulating index is m_1, m_2, m_3------------.So each signal contribute power of P_1, P_2, P_3--------- respectively and $P_r = (m_r^2/2)$ Pc. Let the mew modulating index of AM is k_n so,

$$P_t = (P_1 + P_2 + P_3 + \text{--------}) + P_c$$

Or, $(1 + k_n^2/2)P_c = P_c + P_c/2 \sum_{r=1}^{n} m_r^2$

Or, $P_c + (k_n^2/2) P_c = P_c + P_c/2 \sum_{r=1}^{n} m_r^2$

Or, $k_n^2 = \sum_{r=1}^{n} m_r^2$

$$\therefore k_n = \sqrt{m_1^2 + m_2^2 + m_3^2 + \text{----} m_n^2}$$

This is i.e. k_n is the <u>modulating index</u> of multi tone AM.

MODULATING INDEX Calculation of TONE MODULATION

From the diagram of AM wave, we get

$$E_{max} = (E_c + E_m) \quad \& \quad E_{min} = (E_c - E_m)$$

From the above two we get-

$$E_c = (E_{max} + E_{min})/2 \qquad [\text{Let } Z = E_{min}/E_{max}]$$
$$\& \quad E_m = (E_{max} - E_{min})/2$$
$$\therefore k_a = E_m/E_c = (E_{max} - E_{min})/(E_{max} + E_{min}) = [(1-Z)/(1+Z)]$$

<u>MAXIMUM POWER in AM WAVES</u>

Maximum power in AM waves present when

$$k_a = 1 \text{ (i.e. } E_m = E_c) \text{ So}$$
$$P_{t(max)} = (1 + 1/2)P_c = 1.5P_c$$

DSBSC (Double Side Band Suppressed Carrier)Waveforms

If we deduct the carrier component from AM wave we get DSBSC carrier wave i.e.

$$D(t) = V(t) - \text{carrier}$$
$$= E_c(1 + km(t))\cos \omega_c t - E_c \cos \omega_c t$$
$$= k_a E_c m(t)\cos \omega_c t$$

as $m(t) = \cos \omega_m t$

$$D(t) = k_a E_c/2[\cos(\omega_c + \omega_m)t + \cos(\omega_c - \omega_m)t]$$

So if $D(f) \leftrightarrow D(t)$

Then $D(f) = k_a Ec/2 \ [\delta\{f \pm (f_c + f_m)\} + \delta\{f \pm (f_c - f_m)\}]$

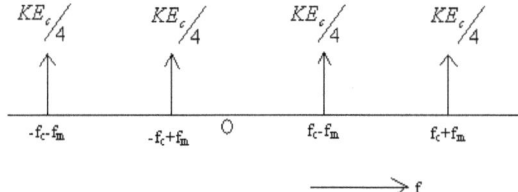

i.e. only two sidebands are present in DSBSC wave.

SSBSC (Single Side Band Suppressed Carrier)

If we suppressed any of the side band from DSBSC we get SSBSC.

So , S(t) = D(t) – upper side band
        = SSBSC of lower side band

or, S(t) = D(t) – lower side band
        = SSBSC of upper side band

i.e. $S(t) = k_aE_c/2[\cos(\omega_c + \omega_m)t]$

or, $S(t) = k_aE_c/2[\cos(\omega_c - \omega_m)t]$

So , $S(f) = k_aEc/2[\delta\{f\pm(f_c + f_m)\}]$ (USB)

Or, $S(f) = k_aEc/2[\delta\{f\pm(f_c - f_m)\}]$ (LSB)

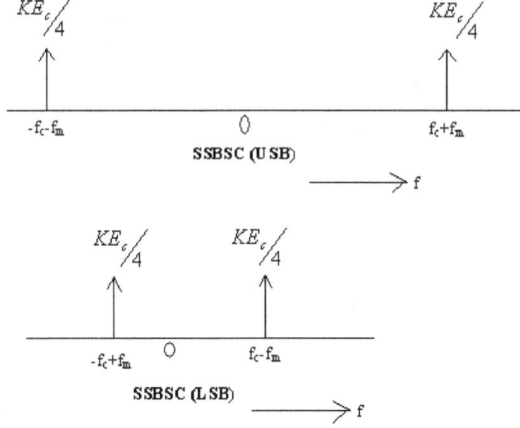

POWER COMPARISION in AM, DSBSC, SSBSC

Power saving in AM to DSBSC:-
$(Saving)_{AM \rightarrow DSB} = (P_{AM} - P_{DSB})/P_{AM}$
$= [(1+k_a^2/2)P_c - (k_a^2/2)P_c]/[(1+k_a^2/2)P_c]$
$= 1/(1 + k_a^2/2)$
This saving is minimum when $(1 + k^2/2)$ is maximum for k = 1.
So then $(1 + k_a^2/2) = (1 + ½) = 3/2$
So, % $(Saving)_{AM \rightarrow DSB}$ minimum = $2/3 \times 100\% = 66.67\%$
∴ If we shift from AM to DSBSC we can save minimum 66.67% power compared to AM.

Power saving in AM to SSBSC:-
$(Saving)_{AM \rightarrow SSB} = (P_{AM} - P_{SSB})/P_{AM}$
$= [(1+ k_a^2/2)P_c - (k_a^2/4)P_c]/[(1+ k_a^2/2)P_c$
$= (1 + k_a^2/4)/(1 + k_a^2/2) = [0.5/(1 + k_a^2/2) + 0.5]$
similarly for k = 1 saving is minimum-
So, % $(Saving)_{AM \rightarrow SSB}$ minimum = $[0.5/(1 +1/2)+ 0.5] \times 100\% = 83.33\%$
∴ If we shift from AM to SSBSC we can save minimum 83.33% power compared to AM.

Power saving in DSBSC to SSBSC:- —
Obviously there is fixed 50% power saved if we shift DSBSC to SSBSC compared to DSBSC
Proof:-
% $(Saving)_{DSB \rightarrow SSB} = (P_{DSB} - P_{SSB})/P_{DSB}$
$= [(k_a^2/2 \ P_c - k_a^2/4 \ P_c)/ k_a^2/2 \ P_c] \times 100\%$
$= [(k_a^2/4)P_c/(k_a^2/2)P_c] \times 100\% = ½ \times 100\% = 50\%(fixed)$
(proved)

Generating scheme for AM, DSBSC & SSBSC

Generations of AM & DSBSC are quite similar but SSBSC generation is far more complex. First we shall discuss the AM & DSBSC generation:-
DEVICE type:-
Two types of devices are used to generate AM & DSBSC----------------------------
1. Gradual Non-linear device or SQUARTING CIRCUIT
2. Pice-wise linear circuit or SWITCHING CIRCUIT

1. NON-LINEAR DEVICE / SQUARE LAW MODULATOR

AM Generation:- Equation of NON-LINEAR device is $V_0 = aV_i + bV_i^2$ (where a, b are constant and V_i = input voltage & V_o = output voltage) .It is usually semiconductor Diode or Transistor and the important thing is ---HERE AMPLITUDE of either CARRIER or MESSAGE is not IMPORTANT.
Now input voltage $V_i = E_c \cos \omega_c t + m (t)$

Analog Communication Engineering Fundamentals

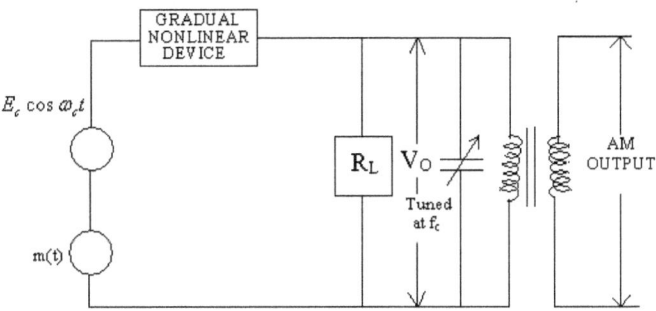

Now $Vo = a\ [E_c \cos \omega_c t + m(t)] + b[E_c \cos \omega_c t + m(t)]^2$

$\qquad = \underline{aE_c \cos \omega_c t} + am\ (t) + bEc^2 \cos^2 \omega_c t + bm^2\ (t) + \underline{2bE_c \cos \omega_c t.m(t)}$

Now tuned transformer which is tuned as f_c, select the underlined forms from $V_0 \rightarrow$ so

$\qquad V\ (t) = E_c \cos \omega_c t + 2bEc \cos \omega_c t.m\ (t)$

$\qquad\qquad = a\ E_c\ [1 + (2b/a).m\ (t)] \cos \omega_c t$

This is the desired AM OUTPUT with $k_a = 2b/a$.

DSBSC Generation:-

a) Balanced Modulator- The above shown AM modulator can be used to obtain DSBSC waves as described below-

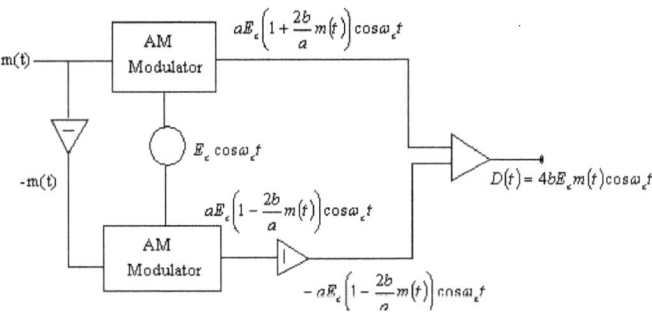

So the output of the above scheme is $D(t)$

and $D(t) = 4b\ E_c\ m(t) \cos \omega_c t$

$\qquad$ which is our desired DSBSC waves.

b) <u>Non-linear DSBSC Modulator</u>- Instead of using 2 AM modulator to generate DSBSC we can directly use two non-linear element say DIODE and a BPF (centre frequency f_c, $f_h = f_c - f_m$, $f_l = f_c - f_m$)-

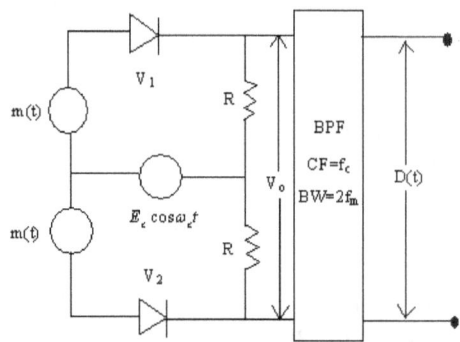

Now, $V1 = a [m(t) + E_c \cos \omega_c t] + b [m(t) + E_c \cos \omega_c t]^2$

and $V_2 = a [-m(t) + E_c \cos \omega_c t] + b [-m(t) + E_c \cos \omega_c t]^2$

Now, $V_O = V_1 + V_2$

So, $V_O = 2 a m(t) + 4bE_c m(t) \cos w_c t$

Now BPF will reject the first term. So,

$$D(t) = 4bE_c m(t) \cos w c t$$

which is or desire DSBSC waves.

2. PICEWISE LINEAR MODULATION or SWITCHING MODULATOR

Consider a PULSE train of $T = T_c = 1/f_c$ (where f_c = carrier frequency) with duty cycle and unit amplitude-

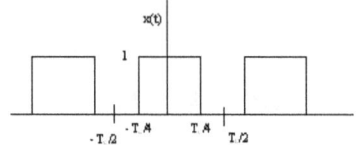

So $X(t) = a_0/T_0 + 2/T_0 \sum_{k=1}^{\infty} (a_k \cos k\omega_0 t + b_k \sin k\omega_0 t)$

$$= a_0/T_c + 2/T_c \sum_{k=1}^{\infty} (a_k \cos k\omega_c t + b_k \sin k\omega_c t)$$

now $a_0/T_c = 1/T_c \int_{-Tc/4}^{+Tc/4} dt = 1/T_c (T_c/4 + T_c/4)$

$$= 1/T_c \times 2T_c/4 = 1/2$$

$\therefore a_0/Tc = 1/2$

Analog Communication Engineering Fundamentals

now as the underline{pulse train} is an underline{even} function of time

$b_k = 0$, now $a_k = 2 \int_0^{Tc/4} 1 \cdot \cos k\omega_c t \, dt$ (due to even function)

$$= (2/k_a \omega_c) \sin \omega_c t \Big|_0^{T0/4}$$
$$= T_c / \pi k_a \sin \frac{k_a \pi}{2}$$
$$So, 2/T_c \; a_k = 2/\pi k_a \sin \frac{k_a \pi}{2}$$

Hence $X(t) = 1/2 + \sum_{k=1}^{\infty} 2/\pi k_a \sin \frac{k_a \pi}{2} \cos k\omega_c t$

$$= 1/2 + 2/\pi (\cos \omega_c t - 1/3 \cos 3\omega_c t + 1/5 \sin 5\omega_c t - \ldots\ldots)$$

let $X(t) = S(t)$

So $S(t) = 1/2 + 2/\pi (\cos \omega_c t - 1/3 \cos 3\omega_c t + 1/5 \sin 5\omega_c t - \ldots\ldots)$

Let $s(t) = 2S(t) - 1$

$\therefore s(t) = 4/\pi (\cos \omega_c t - 1/3 \cos 3\omega_c t + 1/5 \sin 5\omega_c t - \ldots\ldots)$

Here we using DIODE as a SWITCHING DEVICE (Not as non-linear device, as we see carrier) which will 'OPEN' or forward biased when $E_c \rangle 0$ and CLOSED when $E_c \langle 0$. To ensure that E_c should be far greater than E_m i.e. $E_c \rangle\rangle E_m$.

So, here AMPLITUDE OF CARRIER and MODULATING FREQUENCY is most important thing to be considered. So now if we use Ec as a switching pulse pulse S(t) becomes—

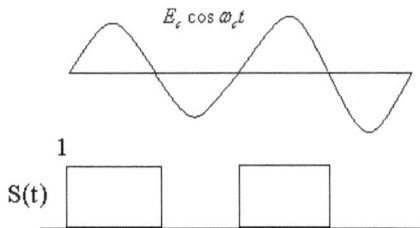

AM GENERATION
The following circuit will generate AM wave-

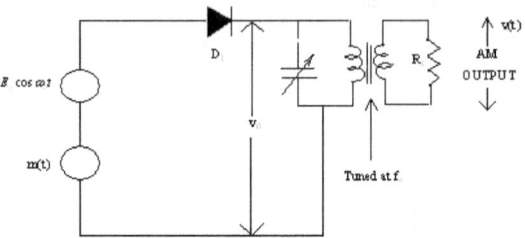

Diode D_1 will be forward biased when $E_c \rangle 0$ and input signal can be pass through the diode, and when $E_c \langle 0$, D_1 in reverse biased state. So diode blocks the input signal. Now this switching pulse of 50% duty cycle with time period $T_c = 1/f_c$ clearly S (t) as we previously described. So the output of the circuit is----------------

$V_o = V_{input} \times S(t)$

$\quad = [E_c \cos \omega_c t + m(t)] \times S(t)$

$\quad = [E_c \cos \omega_c t + m(t)] [1/2 + 2/\pi(\cos \omega_c t - 1/3\cos 3\omega_c t + 1/5\sin 5\omega_c t - ....)]$

$\quad = 2/\pi \ m(t)\cos \omega_c t + E_c/2 \cos \omega_c t +$ other high frequency term.

Now a tuned circuit at fc will select the first two terms. So the output at transformer is (or a BPF of carrier frequency = f_c & BW = $2f_m$)

$\quad V(t) = 2/\pi \ m(t)\cos \omega_c t + E_c/2 \cos \omega_c t$

$\qquad = E_c/2 [1 + (4/\pi \ E_c)m(t)] \cos \omega_c t$

which is desired AM wave with $k_a = 4/\pi \ E_c$

DSBSC GENERATION

If we intended to use the above switching function i.e. S(t) we get two circuits-
1. SHUNT BRIDGE DIODE MODULATOR
2. SERISE BRIDGE DIODE MODULATOR

and if we use S(t) as switching function we get-
3. RING MODULATOR

unlike AM in those schemes for DSBSC generator .
m(t) is the only input and carrier is only used for switching purpose.

1. SHUNT BRIDGE DIODE MODULATOR

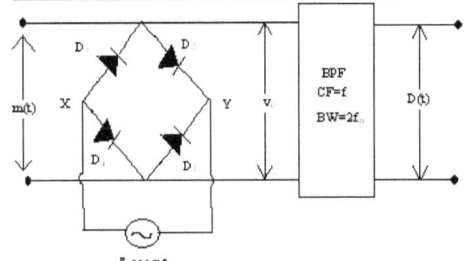

$B \cos \omega t$

When $E_c \rangle 0$, $V_x \rangle V_y$, all diodes (D_1,D_2,D_3,D_4) are forward biased so shorting the input terminals of BPF and restrict the input to reach at BPF and when Ec<0 , $V_y>V_x$ all diodes are reverse biased and then input can reach to BPF. This switching action of diodes already is S (t) and so-

$V_o = V_{IN} \times$ S (t) = m (t) × S (t)

$= m(t) \times [1/2 + 2/\pi(\cos \omega_c t - 1/3 \cos 3\omega_c t + 1/5 \sin 5\omega_c t - ....)]$

$= \frac{1}{2}$ m (t) + 2/ π m(t) cos $\omega_c t$ + other highest order frequency term.

The BPF will select the second term of V_O, So----

D(t) = 2/ π m(t) cos $\omega_c t$

It's our desired DSBSC waves.

2. SERIES BRIDGE DIODE MODULATOR

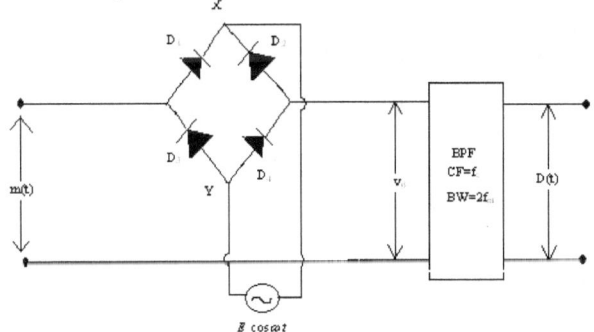

$B \cos \omega t$

Here Diode Bridge connected with series with input. When $E_c>0$, $V_y>V_x$ and all diodes are forward biased and input can reach to BPF. When $E_c<0$, $V_x>V_y$ all diodes are reversed and input cannot reach at BPF. All other condition remains same. The output   D (t) is our desired DSBSC waves.

3.RING MODULATOR

Now consider S(t) $\rightarrow$

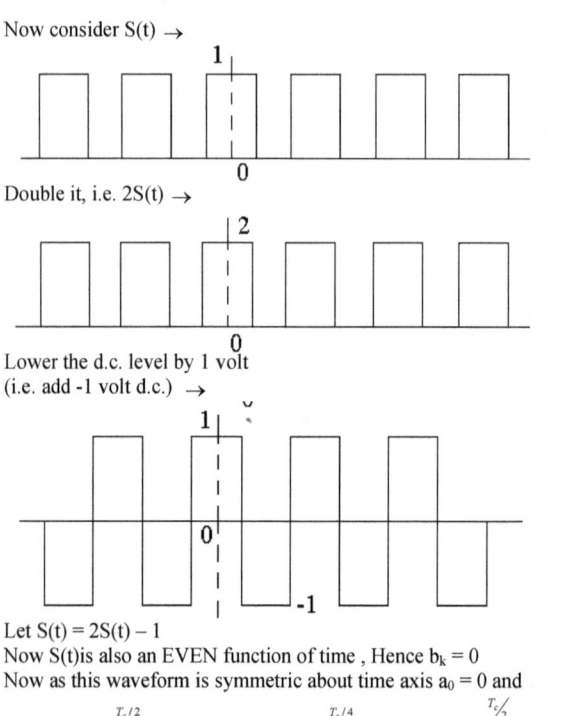

Double it, i.e. 2S(t) $\rightarrow$

Lower the d.c. level by 1 volt
(i.e. add -1 volt d.c.) $\rightarrow$

Let S(t) = 2S(t) – 1
Now S(t)is also an EVEN function of time , Hence $b_k = 0$
Now as this waveform is symmetric about time axis $a_0 = 0$ and

$$a_k = 2 \int_0^{T_c/2} S(t) \cos k\omega_0 t \, dt = 2[\int_0^{T_c/4} \cos k\omega_0 t \, dt - \int_{-T_c/4}^{T_c/2} \cos k\omega_0 t \, dt]$$

$$= 2/k_a \omega_c [\sin (k_a \pi/2) - \sin k_a \pi + \sin(k_a \pi/2)]$$

$$= 4/ k_a \omega_c \sin (k_a \pi/2) = =[4T_c/ k_a (2\pi)] \sin (k_a \pi/2) = [2T_c/ k_a \pi] \sin (k_a \pi/2)$$

so $2/T_c \, a_k = [4/ k_a \pi] \sin (k_a \pi/2)$

Hence, S(t) = $4/\pi [\cos \omega_c t - 1/3 \cos \omega_c t + 1/5 \cos \omega_c t - + \text{----------------}]$

Analog Communication Engineering Fundamentals

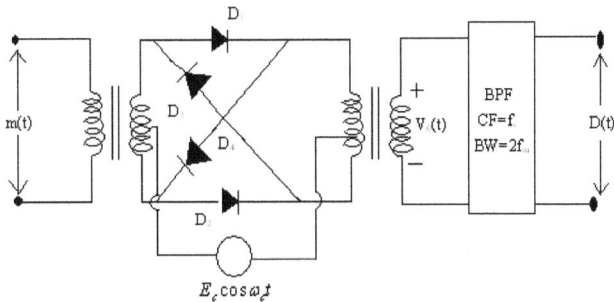

$$E_c \cos \omega_c t$$

When $E_c > 0$ D_1 & D_2 are in forward bias and $V_0(t) = m(t)$. But when $E_c < 0$ D_3 & D_4 are forward bias and $V_0(t) = -m(t)$

So $V_0 = V_{in} \times S(t)$

   $= m(t) \times 4/\pi \; [\cos \omega_c t - 1/3 \cos \omega_c t + 1/5 \cos \omega_c t - + \text{----------------}]$

   $= 4/\pi \; m(t) \cos \omega_c t - +$ other highest order frequency terms.

BPF will select the first term of V_0--------

   So $D(t) = 4/\pi \; m(t) \cos \omega_c t$

Which is our desired DSBSC wave.The above circuit is called RING MODULATOR.

SSBSC GENERATION

1. FILTER METHOD
   If we filter out two sideband (USB & LSB) ————

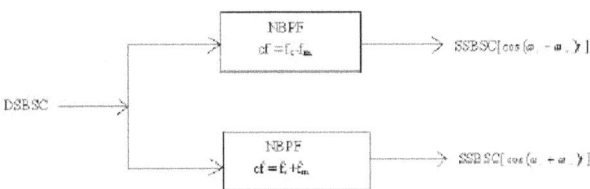

From a DSBSC waves then we can easily produce SSBSC signal (i.e. S(t))

2. PHASE DISCRIMINATION METHOD
   90^0 method

We know DSBSC,$D(t)=m(t) \cos \omega_c t = \cos \omega_m t \cos \omega_c t$

Let 90^0 phase shifted version of m(t)= $\hat{m}(t) = \sin \omega_m t$

and 90^0 phase shifted version of carrier =$\sin \omega_c t$

Now let D'(t)= $\hat{m}(t)\sin \omega_m t = \sin \omega_m t \sin \omega_c t$

Now $D(t)+D'(t)=\cos(\omega_c-\omega_m)t=S(t)$,   LSB

and $D(t)-D'(t)=\cos(\omega_c+\omega_m)t=S(t)$,   USB

The entire scheme is shown below:-

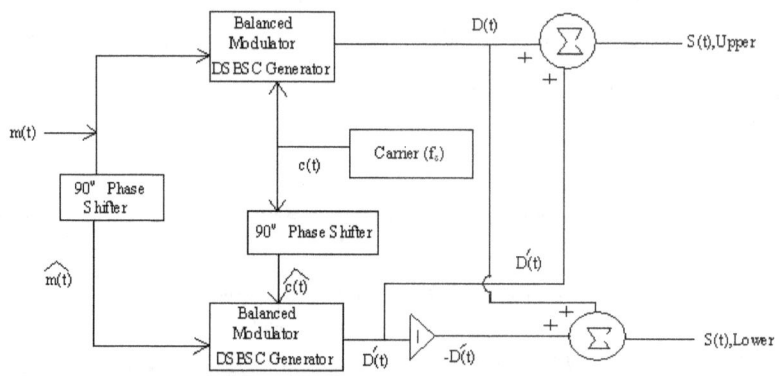

$\alpha-\beta\,Method$:-

Let $\alpha-\beta=90°$

Now $\alpha°$ shifted version of m(t) $=\cos(\omega_m t+\alpha)=m_\alpha(t)$

$\beta°$ shifted version of m(t) $=\cos(\omega_m t+\beta)=m_\beta(t)$

now $m_\alpha(t)\cos\omega_c t=\cos(\omega_m t+\alpha)\cos\omega_c t=\cos\omega_x t\,\cos\omega_c t$

and $m_\beta(t)\sin\omega_c t=\sin\omega_c t\,\cos(\omega_m t+\beta)$

$$=\sin\omega_c t\cos(\omega_z t)$$
$$=\sin\omega_c t\,\cos(\omega_x t-90°)$$
$$=\sin\omega_c t\,\sin\omega_x t$$

Let $D_1(t)=m_\alpha(t)\cos\omega_c t$

& $D_2(t)=m_\beta(t)\sin\omega_c t$

Now $D_1+D_2=\sin\omega_c t\sin\omega_x t+\cos\omega_x t\,\cos\omega_c t=\cos(\omega_c-\omega_x)t$

and $D_1-D_2=\cos\omega_x t\,\cos\omega_c t-\sin\omega_c t\sin\omega_x t=\cos(\omega_c+\omega_x)t$

Now if we shift the phase of (D_1+D_2) by α and of (D_1-D_2) by $-\alpha$ we get-

$S_1(t)=\cos(\omega_c t-\omega_x t+\alpha)=\cos(\omega_c t-\omega_m t)=\cos(\omega_c-\omega_m)t$

& $S_2(t)=\cos(\omega_c t+\omega_x t-\alpha)=\cos(\omega_c t+\omega_m t)=\cos(\omega_c+\omega_m)t$

which are both our desired LSB & USB SSBSC waveforms.

THIRD (WEIVER'S METHOD) METHOD:-

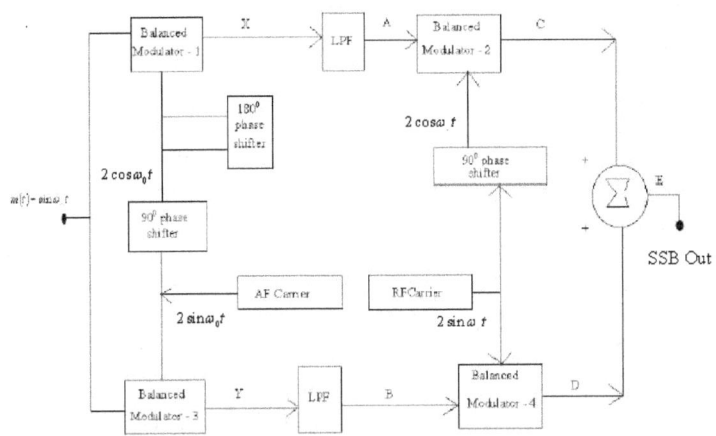

The advantages of PHASE SHIFT method over Filter method that it do not require critical NBPF filter with sharp cut-off which is very difficult to obtain, but phase shift method uses a 90° AF phase shifter , and which is also difficult to obtain a constant 90° phase shifter in all over audio frequency range, so a THIRD method is adopted in the Weiver's method we are using a AF carrier ($\sin \omega_0 t$) to modulate the message first instead of modulating it directly with RF carrier but this system although preserve the advantages of PHASE SHIFT method and can generate SSBSC at any frequency and use of low audio frequency, it is very complex and expensive system compare to FILTER method hence it is rarely used.

DETECTION OF AM, DSBSC & SSBSC

Now we have generated all types of amplitude modulated signals; now we have to demodulate or detected it.

Demodulation or detection is the process of recovering the intelligence contained in the modulated carrier. It is the REVERSE of the modulation process. Modulation involves frequency translation of the information signal to the desired operating range, demodulation involves processing of the translated information signal to shift it back to the original frequency spectrum.

The difference between detection and demodulation is subtle and virtually indicate the same process. If the recovery of information signal involves the reintroduction of the carrier either internally as a part of the applied modulated signal or externally as produced by a local oscillator and through converter action sum and difference frequency terms are produced then the process is technically demodulation.

DSBSC & SSBSC signal both require a demodulation process (Product Law Demodulation) where AM wave require by a Detection process (Square Law Detector, Envelope Detector, Rectifier Detector).

AM DETECTCTION
SQUARE LAW DETECTION:-

1)We know for a squaring circuit

$$V_0 = k_1 V_i + k_2 V_i^2$$

V_i = input , k_1, k_2 = constant

Now let $V_i = v(t) = E_c[1 + k_a m(t)] \cos \omega_c t$ = AM wave

So $V_0(t) = k_1 [E_c \cos \omega_c t + k_a m(t) \cos \omega_c t] + k_2 [E_c^2 + 2k_a E_c^2 m(t) + k_a^2 m^2(t)] \cos^2 \omega_c t$

$= k_1 E_c \cos \omega_c t + k_1 k_a m(t) \cos \omega_c t + k_2 E_c^2 \cos^2 \omega_c t + 2 k_a k_2 E_c^2 m(t) \cos^2 \omega_c t + k_2 E_c^2 k_a^2 m^2(t) \cos^2 \omega_c t$

[now $\cos^2 \omega_c t = 0.5 \cos 2 \omega_c t + 0.5$]

$= k_1 E_c \cos \omega_c t + k_1 k_a m(t) \cos \omega_c t + \underline{k_2 E_c^2 \ 0.5} + k_2 E_c^2 \cos 2 \omega_c t \ 0.5 + \underline{k_a k_2 E_c^2 m(t)}$
$+ k_a k_2 E_c^2 m(t) \cos 2 \omega_c t + k_2 E_c^2 k_a^2 m^2(t) \cos 2 \omega_c t \ . \ 0.5 + 0.5 \ k_2 E_c^2 k_a^2 m^2(t)$

now if we pass V_0 through a LPF it will select underlined terms and suppressed all other terms so-

$V_{LPF} = k_a k_2 E_c^2 m (t) + k_2 E_c^2 \ 0.5$

Now if we pass V_{LPF} through a capacitor it will suppressed D.C term and at output we will get m(t) i.e. the message signal ($V_{OUT} = k_a k_2 E_c^2 m (t) = km(t)$)

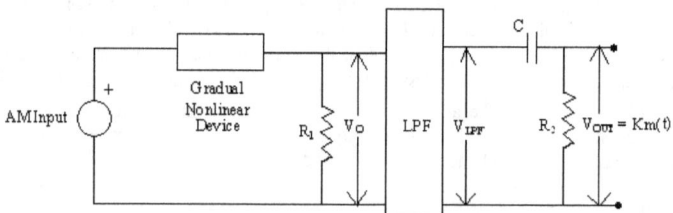

2) Rectifier detector :-

This type of detector is similar to SWITCHING Modulator ,use S(t) to detect m(t).Here again DIODE used as a switching device and Ec>>Em.Now when Ec>0 of input AM wave diode is forward biased, and when Ec<0 the diode is reverse biased. This switching action is clearly the S(t)

Analog Communication Engineering Fundamentals

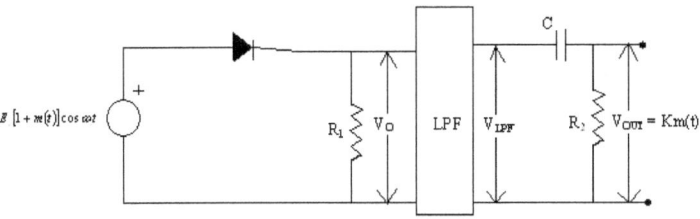

So – V_0 = [AM signal]×S(t)

$$= Ec(1+K_a m(t))\cos \omega_c t \times [\frac{1}{2}+\frac{2}{\pi}(\cos \omega_c t) -\frac{1}{3}\cos 3\omega_c t +\frac{1}{5}\cos 5\omega_c t -\ldots\ldots\ldots)]$$

$$=\frac{1}{2}E_c(1+K_a m(t))\cos \omega_c t +\frac{2}{\pi}E_c(1+K_a m(t))\cos^2 \omega_c t + \text{other terms of higher frequencies.}$$

$$=\frac{1}{2}E_c(1+K_a m(t))\cos \omega_c t +\frac{2}{\pi}E_c(1+K_a m(t))(0.5+0.5\cos 2\omega_c t) + \text{other HF terms.}$$

$$=\frac{1}{2}E_c(1+K_a m(t))\cos \omega_c t +\frac{1}{\pi}E_c(1+K_a m(t)) +\frac{E_c}{\pi}[1+K_a m(t)]\cos 2\omega_c t + \text{other HF terms.}$$

Now a LPF will select the underlined term only , So-----

$$V_{LPF}=\frac{1}{\pi}E_c[1+m(t)]$$

Now if we pass V_{LPF} through a capacitor it will block the DC term and in the output we will get
$V_{out} = \frac{E_c}{\pi}m(t) = Km(t)$ which is the desired message signal.

Envelope detector:

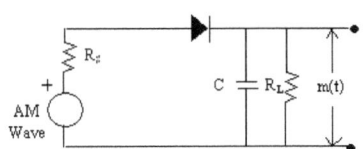

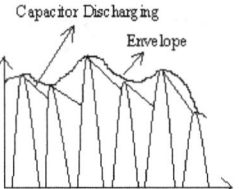

Capacitor Discharging

Envelope

Charging time constant of capacitor =CR_s
Discharging time constant of capacitor =CR_L

If we apply AM wave as a input to the above circuit capacitor C charge up to the peak ,then decaying from peak with time constant CR_L;Again charge up to next peak and so on .Diode will block the negative portion of AM wave ,thus the voltage across capacitor is a REPLICA of m(t),that is ENVELOPE is detected .So it is called Envelope detector .Now to ensure proper operation of the above circuit three condition must be satisfied:

1. $CR_S << \dfrac{1}{f_c}$

That is charging time constant is much less than carrier time period to ensure quickest following of envelope.

2. $CR_L >> \dfrac{1}{f_c}$

That is discharging time constant is much greater then carrier time period to ensure very small decay in the capacitor voltage between two successive peak.

3. The <u>Decaying</u> rate of capacitor voltage should be greater than <u>Rising</u> rate of the successive (next)envelope; Otherwise detectors are not able to trace the next peak. So——$\approx\approx$

We know voltage at capacitor $V_c = E\, e^{-t/R_L C}$

Now from Taylor series expansion we know

$$e^x = 1 + x + -\frac{x^2}{2!} + \frac{x^3}{3!} + \dots\dots\dots$$

taking first two terms--------

$$e^{-t/R_L C} \approx 1 - t/R_L C$$

$$\therefore V_c = E(1 - t/R_L C)$$

Now the slope of the discharging capacitor voltage $dV_c/dt = -E/R_l C$

Now $E = E(t) =$ envelope of AM wave $= E_c(1 + K_a \cos \omega_m t)$

So slope of the envelope $\dfrac{dE}{dt} = -K_a E_c \omega_m \sin \omega_m t$

Now for proper operation of envelope detector------

$$\left|\frac{dV_c}{dt}\right| \geq \left|\frac{dE}{dt}\right|$$

or $E/CR_L \geq K_a E_c \omega_m \sin \omega_m t$

now $E = E_c (1 + K_a \cos \omega_m t)$

so $1/CR_L \geq (K_a \omega_m \sin \omega_m t)/(1 + K_a \cos \omega_m t)$

$\therefore CR_L \leq (1 + K_a \cos \omega_m t)/(K_a \omega_m \sin \omega_m t)$------------(1)

now we should be ready for the worst case. The worst case appear when RHS of (1) takes its minimum value i.e. $\dfrac{d}{dt}[(1 + K_a \cos \omega_m t)/(K_a \omega_m \sin \omega_m t)] = 0$

or $\cos \omega_m t = -K_a$ -------------------(2)

and $\sin \omega_m t = \sqrt{1 - K_a^2}$ --------------(3)

setting condition (2) &(3) in equation (1) we get the worst case--------

$$CR_L \leq \frac{1-K_a^2}{K_a\omega_m\sqrt{1-K_a^2}}$$

$$\therefore CR_L \leq \frac{\sqrt{1-K_a^2}}{K_a\omega_m}$$

the above situation i.e. the missing of negative peak of envelope or successive peak due to very high time constant of R_LC is known as DIAGONAL CLIPPING. To avoid diagonal clipping above condition should be satisfied

we have $CR_L \leq \dfrac{\sqrt{1-K_a^2}}{K_a\omega_m}$

or $C^2R_L^2 \leq \dfrac{1-K_a^2}{K_a^2\omega_m^2}$

or $K_a^2\omega_m^2 C^2 R_L^2 \leq 1 - K_a^2$

or $K_a^2(1+\omega_m^2 C^2 R_L^2) \leq 1$

or $K_a^2 \leq 1/(1+\omega_m^2 C^2 R_L^2)$

$$\therefore K_a \leq \frac{1}{\sqrt{1+\omega_m^2 C^2 R_L^2}}$$

hence the maximum permissible value of modulation index for an AM wave to avoid diagonal clipping in the envelope detection is $\dfrac{1}{\sqrt{1+\omega_m^2 C^2 R_L^2}}$

DEMODULATION of DSBSC & SSBSC:-

We know $D(t) = m(t)\cos\omega_c t$

So $D^2(t) = m^2(t)\cos^2\omega_c t$

Now if we applied D(t) to a squaring circuit output will be $V_0 = k_1 D(t) + k_2 D^2(t)$

Or, $V_0 = k_1m(t)\cos\omega_c t + k_2 m^2(t)\cos^2\omega_c t$

$= k_1m(t)\cos\omega_c t + 0.5 k_2 m^2(t) + 0.5 k_2 m^2(t)\cos2\omega_c t$

we note there is no m(t) in V_0, so can't filter out m(t) from V_0.

Now again $S(t) = \cos(\omega_c \pm \omega_m)t$

So $S^2(t) = \cos^2(\omega_c \pm \omega_m)t$

Now as S(t) input output of a squaring circuit is –

$V_0 = k_1\cos(\omega_c \pm \omega_m)t + k_2[0.5\cos2(\omega_c \pm \omega_m)t + 0.5]$

$= k_1\cos(\omega_c \pm \omega_m)t + k_20.5 + 0.5k_2\cos2(\omega_c \pm \omega_m)t$

Here again we note the absence of m(t) in V_0.

So in conclusion we can say DSBSC & SSBSC can't DEMODULATED by SQUARING CIRCUIT.

PRODUCT DEMODULATOR

(COHERENT/SYNCRONOUS/HOMODYNE DETECTION):-

All diode based modulator discussed previously are called PRODUCT MODULATOR because it multiplies by two input signal. We can use the same circuit with input as DSBSC or SSBSC wave and locally generated carrier which create S(t) by switching operation . Another change we have to make- in PRODUCT MODULATOR we have used a BPF (center frequency ω_c , BW $2\omega_m$) to extract DSBSC and or one LPF (one has $f_h = f_c$) and another HPF ($h_l = f_c$) to extract two SSBSC. But in PRODUCT DEMODULATOR we will use a LPF ($f_H = B$, $f_m <B$ but $B<< f_c$) to extract m(t)-

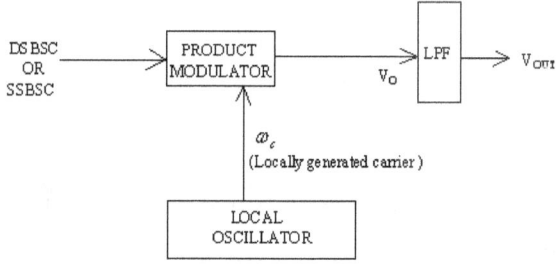

DSBSC Demodulation:-
Let the input into PRODUCT DEMODULATOR is DSBSC wave , then-

$V_0 = D(t) \times S(t)$

$= [m(t) \cos\omega_c t] \times [1/2 +2/\pi (\cos\omega_c t - 1/3 \cos3\,\omega_c t + 1/5 \cos5\,\omega_c t\text{-------})]$

or, $V_0 = \frac{1}{2} m(t) \cos\omega_c t + 2/\pi\, m(t) \cos^2\omega_c t +$ other higher frequency terms

now $2 \cos^2\omega_c t = \cos2\,\omega_c t + 1$

so, $V_0 = \frac{1}{2} m(t) \cos\omega_c t + 1/\pi\, m(t) \cos2\,\omega_c t + 1/\pi\, m(t) +$ other higher frequency terms

Simply LPF of $f_h = B$ where $B>f_m$ but $B << f_c$ will select the underlined term which is message itself i.e. $V_{OUT} = km(t)$

In this way we can demodulate the DSBSC waveform using PRODUCT DEMODULATOR.

QUADRATURE NULL EFFECT:-

Let consider a phase shift locally generated carrier with original carrier i.e. the locally generated carrier is-

$C(t) = E_c \cos (\omega_c t + \theta)$ θ = phase difference with original carrier at transmitter (modulator) so replacing $\omega_c t$ by ($\omega_c t + \theta$) in S(t) , we get-

$V_0 = \frac{1}{2} m(t) \cos\omega_c t + 2/\pi\, m(t) \cos(\omega_c t + \theta) +$ other $\cos\omega_c t$ HF terms

Now $2\cos(\omega_c t + \theta) \cos \omega_c t = \cos(2\omega_c t + \theta) + \cos\theta$

So $V_0 = \frac{1}{2} m(t) \cos\omega_c t + 1/\pi\, m(t)[\cos(2\omega_c t + \theta) + \cos\theta]$ + other HF terms

Hence obviously $V_{OUT} = km(t) \cos\theta$

So AMPLITUDE OF DEMODULATED MESSAGE is MAXIMUM when there is NO PHASE ERROR between locally generated carrier and original carrier i.e. at $\theta = 0$ now for $\theta = \pm\pi/2$ $\cos\theta = 0$. Hence ZERO signal will occur for $\theta = +\pi/2$ represent the QUADRATURE NULL EFFECR of COHERENT DETECTION.

In practice we would find that 'θ' is varies randomly with time i.e. produce a DISTORTED version of message which is not desirable . So we need to maintain perfect SYNCHRONYZATION between incoming DSBSC's carrier and local oscillator's carrier. That is why product demodulation also called SYNCHRONOUS or COHERENT detection.

SSBSC Demodulation:-

We know SSBSC $= k\cos(\omega_c \pm \omega_m)t$

Now for a product demodulator-

$V_0 = [SSBSC] \times S(t)$ [as SSB is input]

$= k_a \cos(\omega_c \pm \omega_m)t + [1/2 + 2/\pi\,(\cos\omega_c t - 1/3\cos 3\omega_c t + 1/5\cos 5\omega_c t\text{-------})]$

$= k_a/2 \cos(\omega_c \pm \omega_m)t + 2 k_a/\pi\,\cos(\omega_c \pm \omega_m)t \cos\omega_c t$ + other HF terms

$= k_a/2 \cos(\omega_c \pm \omega_m)t + k_a/\pi\,[\cos(2\omega_c \pm \omega_m)t + \cos(\pm\omega_m)t]$ + other HF terms

$= k_a/2 \cos(\omega_c \pm \omega_m)t + k_a/\pi\cos(2\omega_c \pm \omega_m)t + k_a/\pi\,\cos(\omega_m t) +$

other HF terms

[as $\cos(-\theta) = \cos\theta$]

now if we pass V_0 through a LPF the output is information itself i.e. $V_{OUT} =$

$k_a/\pi\cos(\omega_m t)$

$= k_1 m(t)$ = message.

VSBSC
Generation & Detection

Vestigial Sideband Modulation: - For bandwidth conservation we know the best system would be SSBSC, but practical SSBSC have poor low frequency response. The DSBSC works quite well for low message frequencies, but the transmission bandwidth is twice that of SSBSC.Clearly a compromise modulation between SSBSC & DSBSC is dopted.This is called VSBSC.

The VSBSC waveform is obtained from DSBSC by filtering it(or from AM first suppressing carrier component)such a fashion that one sideband is passed almost completely while just a TRACE or VESTIGIAL of the other sideband is included.It must have odd symmetry about the carrier horizontal axis.

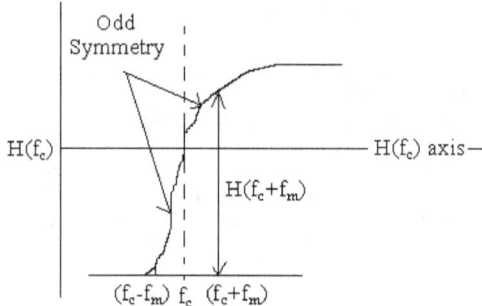

VSBSC SPECTRA

VSBSC-SPECTRA

So VSBSC can be obtained passing DSBSC through a filter which has pass band characteristics as VSBSC Spectra. WHILE THE EXACT SHAPE OF THE RESPONSE IS NOT CRUSIAL; it must have odd symmetry about carrier frequency & a RELATIVE RESPONSE OF ½ AT THAT POINT .So it can be written as $2H(f_c)= H(f_c+f_m)+H(f_c-f_m)$

Now if $H(f_c)=0.5$

then $H(f_c+f_m)=0.5+a$

$H(f_c-f_m)= 0.5-a$

Due to odd symmetry about carrier.

Now if we pass a DSBSC signal through this filter, it will weight (f_c+f_m) component by $H(f_c+f_m)$ & (f_c-f_m) by $H(f_c-f_m)$ .So----

VSBSC=DSBSC×VSBSC

　　　　　　　Filter

$= (2K \cos \omega_c t \cos \omega_m t) \times VSBSC$

　　　　　　　filter

$= K[\ H(f_c+f_m) \cos (\omega_c + \omega_m)t + H(f_c-f_m)\cos(\omega_c - \omega_m)t]$

$= K[(0.5+a)\cos (\omega_c + \omega_m)t + (0.5-a) \cos(\omega_c - \omega_m)t]$

$= K[0.5\{\ \cos (\omega_c + \omega_m)t + \cos(\omega_c - \omega_m)t\}+a\{\ \cos (\omega_c + \omega_m)t - \cos(\omega_c - \omega_m)t\}]$

$= K[\cos \omega_m t \cos \omega_c t - 2a \sin \omega_m t \sin \omega_c t]$

$\therefore VSBSC = V(t) = K[\cos \omega_m t \cos \omega_c t - 2a \sin \omega_m t \sin \omega_c t$

The above expression shows that VSBSC is basically a compromise between DSBSC & SSBSC. If a= 0 i.e. $H(f_c+f_m) = H(f_c-f_m)=0.5$ i.e. both sideband have equal weight the above equation form the the equation of DSBSC and for a=± 0.5 i.e. $H(f_c-f_m)=0$ or $H(f_c+f_m)=0$ i.e suppressing one sideband completely ,the above equation form the equation of SSBSC.

DETECTION OF VSBSC

Detection of VSBSC waveform is same as that of SSBSC & DSBSC.Here we use again the PRODUCT MODULATOR with a LPF.

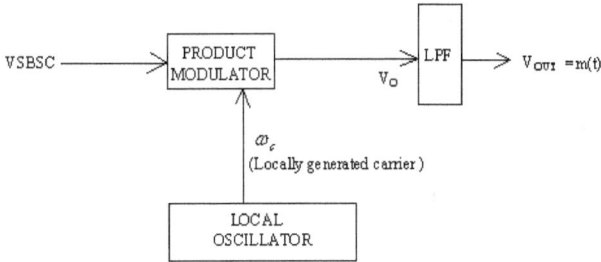

Now $V_0 = V(t) \times S(t)$

Or $V_0 =$

$k_a(\cos \omega_c t \cos \omega_m t - 2a \sin \omega_c t \sin \omega_m t) \times [1/2 + 2/\pi(\cos \omega_c t - 1/3 \cos 3\omega_c t + 1/5 \cos 5\omega_c t - ....)] =$

$1/2 \, k_a(\cos \omega_c t \cos \omega_m t - 2a \sin \omega_c t \sin \omega_m t) + 2/\pi k_a \cos \omega_m t \cos^2 \omega_c t + otherHFvalues$

$= VSBSC + 2/\pi k_a \cos \omega_m t(0.5 + 0.5 \cos 2\omega_c t) + otherHFvalues$

$= VSBSC + k_a/\pi \cos \omega_m t + k_a/\pi \cos \omega_m t \cos 2\omega_c t + otherHFvalues$

$= VSBSC + k_1 m(t) + k_a/\pi \cos \omega_m t \cos 2\omega_c t + other \ HF \ values$

now LPF will select the m(t) form.

V_0, Hence $V_{out} = K_1 m(t)$.This is the message.

Chapter 6

Angle Modulation

FREQUENCY MODULATION

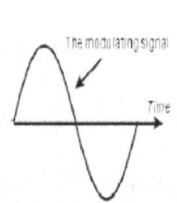

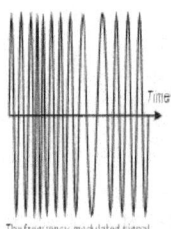

The modulating signal

Time

Time

The frequency-modulated signal

In FM the instantaneous frequency deviation of carrier wave varied directly proportional with the amplitude of message.

Let, instantaneous frequency of carrier wave is ω_i .Hence the deviation

$$\Delta\omega_c = \omega_i - \omega_c = \omega_d E_m(t)$$

$E_m(t)$=message, ω_d =deviation const. in $[RAD] [S]^{-1} [VOLT]^{-1}$

$$\therefore \omega_i = \omega_c + \omega_d E_m(t)$$

Now the FM wave is V_{FM}=Ec cos $\theta(t)$

E_c=amplitude of carrier

$\theta(t)= \int \omega_i \, dt$

Now, $\theta(t) = \int \omega_i \, dt = \int_t [\omega_c + \omega_d E_m(t)] dt$

$$= \omega_c t + \omega_d \int_t E_m(t) dt$$

$$\therefore V_{FM}(t) = \cos[\omega_c t + \omega_d \int_t E_m(t) \, dt \ ]$$

this is the expression of FM wave for MULTI frequency.

TONE modulation:

For a single frequency message i.e.

$E_m(t) = E_m \cos \omega_m t$

Hence $\int_t E_m(t) dt = E_m / \omega_m \sin \omega_m t$

So $V_{Fm}^{tone}(t) = E_c \cos[\omega_c t + \omega_d E_m / \omega_m \sin \omega_m t]$

Now obviously $\omega_d E_m$ =maximum frequency deviation.

ω_m = modulating frequency.

Defining modulating index $m_f = \omega_d E_m / \omega_m$

$V_{Fm}^{tone}(t) = E_c \cos[\omega_c t + m_f \sin \omega_m t]$

m_f = maximum frequency deviation /modulating frequency.

$= \Delta f_{dmax} / f_m$

So we can say m_f is the maximum phase deviation (for $\sin \omega_m t = 1$) of carrier in tone modulation.

MODULATION INDEX in FM is only defined for TONE MODULATION.

As the amplitude of the FM wave is always constant therefore regardless of the message the average power is $Pc = (Ec/\sqrt{2})^2 = 1/2 \ Ec^2$.

MULTI -TONE FM WAVE:

For a message of multi frequency i.e. if m(t) content a group of 'sin' waves of different frequencies which may be completely unrelated or harmonically related for a carrier wave of Ec $\cos \omega_c t$ and if it is modulated by a message content three different frequencies then resulting FM wave is

$F(t) = E_c \cos[\omega_c t + m_{f1} \sin \omega_1 t + m_{f2} \sin \omega_2 t + m_{f3} \sin \omega_3 t \ ]$

Where m_{f1}, m_{f2}, m_{f3} is the modulation index for $1^{st} \ 2^{nd} \ \& \ 3^{rd}$ tone respectively.

TRANSMISSION BAND WIDTH OF FM WAVES

For large values of modulation index m_f the BW approaches and is slightly greater than the total frequency deviation $2\Delta f_{d \, max}$.On the other hand for small values of modulation index m_f the spectrum of the FM wave is effectively limited to the carrier frequency f_c and one pair of side band frequencies at $f_c \pm f_m$. So that BW approaches to $2f_m$ .We may sum up above two cases and an approximate rule for the transmission BW as follows:

$B_T \cong 2\Delta f_{d,max} + 2f_m$

$= 2\Delta f_{d,max}(1 + f_m / \Delta f_{d,max})$

$= 2\Delta f_{d,max}(1 + 1/m_f)$

$= 2f_m(1 + \Delta f_{d,max} / f_m)$

$= 2f_m(1 + m_f)$

$$B_T = 2\Delta f_{d,max} + 2f_m$$
$$= 2\Delta f_{d,max}(1 + f_m / \Delta f_{d,max})$$
$$= 2\Delta f_{d,max}(1 + 1/m_f)$$

This relation is known as CARSON's rule

Analog Communication Engineering Fundamentals

For multi frequency or multitone FM if w is the largest frequency of the FM then instead of the m_f, we calculate D (deviation ratio) $= \Delta f_{d,max} / w$ and replacing m_f by D and f_m by w we get Carson's rule for transmission bandwidth for multitone FM is

$$B_T \cong 2\Delta f_{d,max} + 2w$$
$$= 2\Delta f_{d,max}(1+1/D)$$

Narrow Band FM(NBFM):
We know for tone FM, the FM wave is-
$$V_{fm}(t) = E_c \cos(\omega_c t + m_f \sin \omega_m t)$$
$$= E_c \cos \omega_c t \cos(m_f \sin \omega_m t) - E_c \sin \omega_c t \sin(m_f \sin \omega_m t)$$

Now for a narrowband FM $\Delta f_{d,max}$ is very small. So $m_f = \Delta f_{d,max} / \omega_m$ also very small. So then $\cos(m_f \sin \omega_m t) \approx 1$ and $\sin(m_f \sin \omega_m t) \approx m_f \sin \omega_m t$ (provided that m_f small compare to 1 radian), hence

$$V_{NBFM}(t) = E_c \cos \omega_c t - E_c m_f \sin \omega_m t \sin \omega_c t$$

Now C(t) = carrier frequency $= E_c \cos \omega_c t$

Now $\overline{C} = 90^0$ phase shifted version of carrier
$$= E_c \cos(\omega_c t \pm 90^0) = E_c \sin \omega_c t$$

$\therefore V_{NBFM}(t) = C(t) - \overline{C} \ (m_f \sin \omega_m t)$

now $(m_f \sin \omega_m t) = \omega_d E_m / \omega_m (\sin \omega_m t)$

$$= \omega_d \int_t E_m \cos \omega_m t \, dt$$

$$= \omega_d \int_t m(t) dt$$

m(t) = modulating wave

$\therefore V_{NBFM}(t) = C(t) - \overline{C} \ [\omega_d \int_t m(t) dt]$

So the generation scheme of NBFM is :-

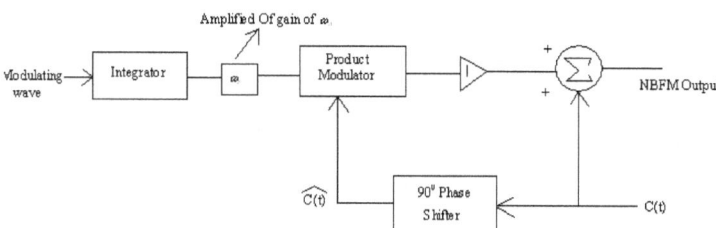

Now we know $B_T \approx \Delta f_{d,max} + 2w$

As $\Delta f_{d,max}$ is very small is NBFM, the transmission bandwidth of NBFM is $B_{T,NBFM} \approx 2w$.

For wideband FM $\Delta f_{d,max} \gg 1$ or $m_f \gg 1$

Hence $B_T \approx 2\Delta f_{d,max}(1+1/m_f)$

$\qquad \approx 2\Delta f_{d,max}$.

For WBFM maximum deviation is 75 kHz.
For NBFM maximum deviation is 5 kHz.
WBFM reduce larger BW as 200 kHz.
(With guard band of 25 kHz each side, BW=B_T+2×25
$\qquad\qquad\qquad\qquad\qquad\qquad = 2 \times 75 + 2 \times 25$
$\qquad\qquad\qquad\qquad\qquad\qquad = 200$ KHz.)
WBFM is used for radio communication.
NBFM is used FM mobile communication services such as police, ambulance etc.

PHASE MODULATION (PM):

In this modulation phase of the carrier signal varied according to instantaneous amplitude of message.
Now $C(t) = E_r \cos(\omega_c t + \varphi)$
$\varphi_{max} = k_p E_m$ (k_p = constant,
$\qquad\qquad E_m$ = maximum amplitude of message)
This φ_{max} is called MODULATION INDEX m_p for phase modulation.
Now for a PM system $\varphi = k_p m(t)$

$$= k_p\, E_m \cos \omega_m t \qquad [m_p = k_p\, E_m]$$
$$= m_p \cos \omega_m t$$

∴ the PM wave is (tone modulation):

$$V_{pm}(t) = E_c \cos(\omega_c t + m_p \sin \omega_m t)$$
$$m_p = \text{modulation index} = k\, E_m \quad (k = \text{constant})$$
$$\text{So} \quad \theta_{pm} = (\omega_c t + m_p \cos \omega_m t)$$

$$\therefore \omega_i = d\theta_{pm}/dt = \omega_c - m_p \omega_m \sin \omega_m t$$

$$\Delta f_{d,\max} = m_p \omega_m \sin \omega_m t$$

$$m_p = \text{modulation index} = k\, E_m \quad (k = \text{const})$$
$$\text{Frequency deviation} \quad \Delta f = k E_m \omega_m \sin \omega_m\, t$$
$$(\Delta f)_{\max} = k E_m \omega_m = m_p \omega_m$$

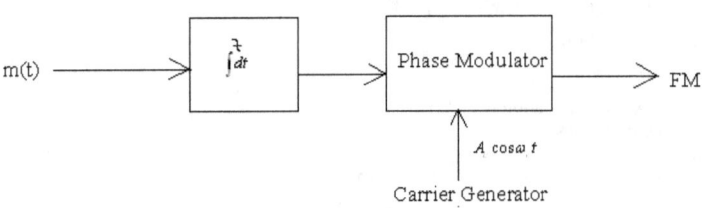

$$\int_t m(t)\,dt = E_m/\omega_m (\sin \omega_m t) = \text{input as a message in PM modulator}$$

So $V_0 = E_c \cos(\omega_c t + k.E_m/\omega_m \sin \omega_m t)$
$$= E_c \cos(\omega_c t + m_f \sin \omega_m t)$$
$$= \text{FM output}$$

Wide Band FM generator:

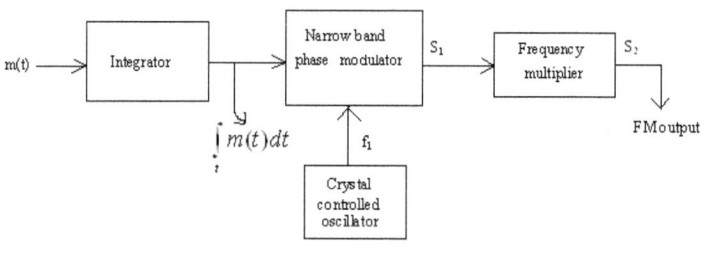

The o/p of phase multiplier is-

$$S_1 = A_1 \cos[2\pi f_1 t + 2\pi k_f \int_t m(t)dt]$$

For Tone modulation

$$S_1 = A_1 \cos[2\pi f_1 t + m_{f_1} \sin \omega_m t]$$

$m_f \leq 0.3$ radian to keep

Now frequency multiplier multiply the freq. of S_1 by n-times

So $S_2 = A_c \cos[2\pi n f_1 t + n m_{f_1} \sin \omega_m t]$

If $m_f = n \, m_{f1}$ and $f_c = n f_1$

The $S_2 = A_c \cos[\omega_c t + m_f \sin \omega_m t]$

= desired FM wave.

Properly choosing value of "n" we can produce or get desired wideband FM.

PM bandwidth:

PM Bandwidths given by Carson's rule-

$$(BW)_{PM} \cong 2 \ (\Delta f)_{max} = 2 k E_m \omega_m .$$

FM GENERATION:

    The previous described circuit is an indirect method of FM generation. Here is a direct method.

VARACTER DIODE CIRCUIT:

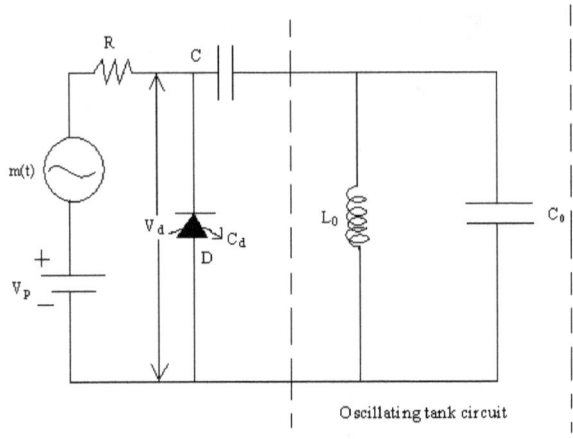

Oscillating tank circuit

Diode D is a varacter diode whose junction capacitance varies with reverse bias voltage applied across in it.

Its junction capacitance $C_d = k (V_d)^{-1/2}$

Capacitance C is very small compared to C_0 and also its reactance kept so high in the highest modulating frequency compare to R for modulating frequency just not by passed through C. $V_p >$ max m(t).

Now voltage across diode $V_d = m(t) + V_p$

So $C_d = k [V_p + m(t)]^{-1/2}$ where V_p is the reverse polarity voltage always kept diode D in reverse bias.

Now as $C << C_0$ the capacitance of the oscillator is $C_t = C_d + C_0$.

Hence instantaneous frequency generated oscillator is $\omega_i = 1/\sqrt{LC_t}$

$$= [L_0(C_0 + kV^{-1/2}{}_d)]^{-1/2}$$

$\omega_i = [L_0 \{C_0 + k(m(t) + V_p)^{-1/2}\}]^{-1/2}$

Where $\omega_c = 1/\sqrt{L_0 C_0}$.

Amplitude of sidebands of FM wave it pair of sidebands of FM wave:

If pair of sidebands represent by n (i.e. carrier represent as n = 0, first pair of sidebands ie component of ($f_c \pm f_m$) represent as n =± 1, second pair of sidebands ie at ($f_c \pm 2f_m$)

Represents n =± 2)

Then the amplitude of each sideband is given by,

Analog Communication Engineering Fundamentals

$$A_c \, Jn \, (m_f) = A_c \sum_{r=0}^{\infty} (-1)^r (0.5 m_f)^{|n|+2r} / r! \, (|n|+r) \, !$$

Also $Jn \, (m_f) = J\text{-}n \, (m_f)$ n even
$Jn \, (m_f) = - \, J\text{-}n \, (m_f)$ n odd

m_f = modulation index . A_c = amplitude of unmodulated carrier

Hence
$$\varphi_{FM}(t) = A_c J_0(m_f) \cos \omega_c t + A_c J_1(m_f)[\cos(\omega_c + \omega_m)t - \cos(\omega_c - \omega_m)t]$$
$$+ \; A_c J_2(m_f)[\cos(\omega_c + 2\omega_m)t - \cos(\omega_c - 2\omega_m)t]$$
$$+ \; A_c J_3(m_f)[\cos(\omega_c + 3\omega_m)t - \cos(\omega_c - 3\omega_m)t]$$
$$+ \ldots\ldots\ldots + \ldots\ldots\ldots\ldots$$

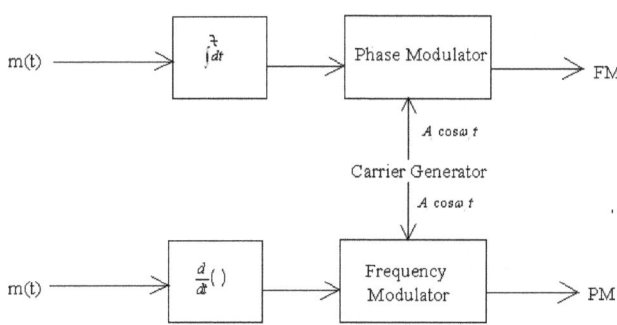

FM DISCRIMINATOR:

Here we discussed a simple FM discriminator called single tuned Discriminator or simple slope detector.

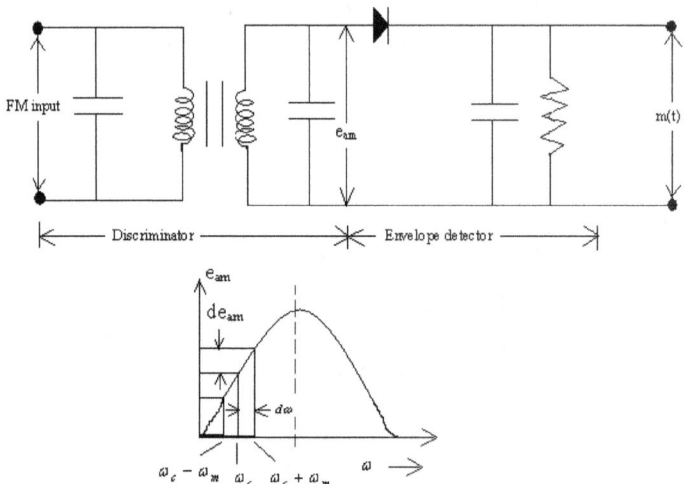

This ckt. Consist of a tuned circuit which is slightly detuned from the carrier frequency ω_c. this ckt. Converts the FM signal into an AM signal then the AM signal is detected by an Envelope detector producing the message m (t) at output. The frequency response of this detuned (slightly off tuned at ω_c) is shown in fig. the slope of the characteristics curve is given as $\alpha = de_{am}/d\omega$. Now a small variation in the frequency takes $\Delta\omega_c$, of the input will produce a change in the amplitude of e_{am} by an amount of $\alpha\,(\Delta\omega_c)$. This freq variation at the input of the discriminator produces amplitude variation at its output. In this way FM signal is converted to a AM signal which is detected by an envelope detector.

Analog Communication Engineering Fundamentals